Lecture Notes in Mathematics

Edited by A. Dold and B. Eckm

577

J. Peter May

E_∞ Ring Spaces and E_∞ Ring Spectra

with contributions by
Frank Quinn, Nigel Ray,
and Jørgen Tornehave

Springer-Verlag
Berlin · Heidelberg · New York 1977

Author
J. Peter May
The University of Chicago
Department of Mathematics
5734 University Avenue
Chicago, IL 60637/USA

AMS Subject Classifications (1970): 18D10, 18F25, 55B15, 55B20, 55D35, 55E50, 55F15, 55F25, 55F35, 55F45, 55F50, 55F60, 55G25

ISBN 3-540-08136-4 Springer-Verlag Berlin · Heidelberg · New York
ISBN 0-387-08136-4 Springer-Verlag New York · Heidelberg · Berlin

Printed in Germany

Printing and binding: Beltz Offsetdruck, Hemsbach/Bergstr. 2141/3140-543210

Contents

Introduction

In topology, there is a dichotomy between two general classes of spaces and ways of thinking about their roles. On the one hand, there are the concrete geometric spaces, most importantly the various types of manifolds. Typical problems one proposes to study about such spaces are their classification, at least up to cobordism, and the obstructions to the existence of an equivalence of a given space with a space with a richer type of structure. Bundles and fibrations over geometric spaces generally play a central role in the solution to such problems. On the other hand, there are the classifying spaces for bundle and fibration theories and other cohomological invariants of spaces. These are thought of as tools for the analysis of geometric problems, and it is a familiar fact that theorems on the classifying space level often translate to yield intrinsic information on the bundle theory level. Thus, for example, Bott periodicity originated as a statement about the homotopy types of classifying spaces, but is most usefully interpreted as a statement about bundles and their tensor products.

The last decade has seen an intensive analysis of the homotopy types of the classifying spaces of geometric topology, with a view towards applications to the classification and obstruction theory of spaces less richly structured than differentiable manifolds. However, very much more structure than the mere homotopy type is revelant. The study of topological and PL manifolds and of Poincare duality spaces forces consideration of bundle and fibration theories the cohomology of whose classifying spaces is wholly inaccessible to the classical techniques and invariants of homotopy theory.

The appropriate framework for the study of these classifying spaces is infinite loop space theory and, in particular, its multiplicative elaboration which is the theme of this book. This is also the appropriate framework for the most structured development of algebraic K-theory, by which we understand the homotopy theory of discrete categories, and one of the main goals of this volume is a complete analysis of the relationship between the classifying spaces of geometric topology and the infinite loop spaces of algebraic K-theory. The results obtained have powerful calculational consequences, which are presented in [26]. For example, they make it possible to pass quite directly from representation theoretical computations of the homologies of appropriate finite groups to detailed analysis of characteristic classes for topological bundles (away from 2) and spherical fibrations.

From the point of view of classical algebraic topology, infinite loop space theory may be thought of as the use of unstable methods for the study of stable homotopy theory. Its starting point is a recognition principle for infinite loop spaces that allows one to pass back and forth between spectra and spaces with appropriate internal structure, namely E_∞ spaces. We shall enrich this additive theory with a multiplicative structure which allows one to pass back and forth between E_∞ ring spectra and E_∞ ring spaces.

Conceptually, the new theory is different in kind from the old: the appropriate multiplicative structure on spectra is itself unstable in that it appears not to admit an equivalent formulation expressible solely in terms of structure visible to the stable homotopy category. This is because the relevant structure requires very precise algebraic data on the point-set level on the spaces which together comprise the spectrum. The applications that concern us here center around space level exploitation of this algebraic data. It has very recently become possible to express a significant portion of the structure in terms of maps in the stable category. This reformulation leads to applications in stable homotopy theory and will be the subject of a future volume.

The prototype of an E_∞ ring spectrum is the sphere spectrum $Q_\infty S^0$. Its zeroth space QS^0 is the prototype of an E_∞ ring space. For purposes of both theory and calculation, the essential feature of QS^0 is the interrelationship between its additive infinite loop space structure and the multiplicative infinite loop space structure on the component SF of its identity element. It is such interrelationships which will be codified in our basic definitions.

If we ignore QS^0 and concentrate on SF, then the interest focuses on geometric topology and, in particular, on the relationships among F, Top, PL, the classical groups, and their homogeneous and classifying spaces. We give consistent infinite loop space structures to these spaces in chapter I by use of the theory of E_∞ spaces developed in [45] (and recapitulated later in this volume). The basic results here were originally due to Boardman and Vogt [19, 20].

Our construction of an infinite loop space structure on SF depends on use not just of the SF(n) but of the SF(V) for all finite-dimensional real inner product spaces $V \subset R^\infty$. If we are to see this structure in the spectrum $Q_\infty S^0$, then spectra too must be indexed on such V rather than just on the non-negative integers. We define such spectra in chapter II, which is mainly an extract and summary from [48] of that slight amount of information about coordinate-free spectra and the stable homotopy category needed in this book.

We reach the definition of E_∞ ring spectra in chapter IV, which is joint work with Frank Quinn and Nigel Ray. The fundamental idea of putting an E_∞ space structure on a spectrum in order to obtain the richest possible notion of a ring spectrum is entirely due to Frank Quinn. A manageable and technically correct way of doing this eluded us for some time. The essential insight leading to the correct definition came from Nigel Ray, who pointed out that good concrete constructions of Thom spectra gave naturally occuring examples of spectra with the right kind of structure and that these examples could be taken as models on which to pattern the general definition.

One key family of classifying spaces in geometric topology is missing from chapter I, namely the classifying spaces for sphere bundles of a given type oriented with respect to a specified cohomology theory. In fact, no concrete constructions of such classifying spaces exist in the literature. We remedy this in chapter III by use of the general classification theory for fibrations and bundles developed in [47]. In §3 of chapter IV, we give these classifying spaces infinite loop space structures when the specified cohomology theory is represented by an E_∞ ring spectrum.

In chapter V, we demonstrate that formal analysis on the classifying space level allows one to deduce sharpened versions of the results of Adams [4,5] on J(X) and of Sullivan [72] on topological bundle theory (away from the prime 2) from the kO-orientation of Spin-bundles and the kO[1/2]-orientation of STop-bundles together with the Adams operations, the cannibalistic classes derived from them by use of the specified orientations, and the Adams conjecture. In the last section, we combine various results from throughout this volume with results of Adams and Priddy [8], Madsen, Snaith, and Tornehave [42], and Ligaard [38] to analyze the infinite loop space structure of BTop (away from 2) and of various other classifying spaces utilized in earlier parts of the chapter. This material completes most of the program envisioned in a preprint version of this chapter.

Aside from its last section, chapter V is largely independent of infinite loop space theory and is reasonably self-contained. However, its earlier sections do make essential use of the fact that real connective K-theory is represented by an E_∞ ring spectrum kO. This fact implies in particular that $BO_\otimes$, the component of the identity element of the zeroth space of kO, is an infinite loop space. Much more significantly, it encodes the interrelationship between kO and the spectrum determined by $BO_{\otimes}$. It is one easy consequence of this interrelationship which is exploited in chapter V. In contrast to the case of Thom spectra, kO does not seem to occur in nature as an E_∞ ring spectrum and must therefore be manufactured.

This brings us to the last four chapters which, aside from use of definitions contained in the first sections of chapters I, II, and IV, are largely independent of the first five. In chapter VI, we define E_∞ ring spaces and show that the classifying spaces of categories with appropriate internal structure, namely bipermutative categories, are examples of such spaces. In chapter VII, we retool the machine constructed in [45 and 46] for the manufacture of spectra from E_∞ spaces so as to make it turn out coordinate-free spectra. We then show that if the machine is fed the additive structure of an E_∞ ring space, then it turns out an E_∞ ring spectrum. One immediate application is a multiplicatively enriched version of the Barratt-Quillen theorem [16,68,46] to the effect that QS^0, and thus the stable homotopy groups of spheres, can be constructed out of symmetric groups. Our version shows that the infinite loop space SF, and thus the classifying space BSF for stable spherical fibrations, can also be constructed out of symmetric groups.

In the last section of chapter VII, we study the spectra turned out when the machine is fed the multiplicative structure of an E_∞ ring space. In particular, we obtain a purely multiplicative version of the relationship between SF and symmetric groups. In chapter VIII, we give the promised analysis of the relationship between the classifying spaces of geometric topology and the infinite loop spaces of algebraic K-theory. This basic material is a mosaic of results due to Jørgen Tornehave and myself and includes new proofs and generalizations of the results originally given in his thesis [75] and in his unpublished preprints

[76] and [77]; it is presented here under joint authorship. The connection between algebraic and topological K-theory was established in the work of Quillen [58,59], and we show that the maps given by Brauer lifting which he used to prove the Adams conjecture are infinite loop maps, both additively and multiplicatively. Via the Frobenius automorphism, this information yields a good understanding of the infinite loop space BCokerJ, which is the basic building block for BSF and for BTop (away from 2) and turns out to be the classifying space for j-oriented stable spherical fibrations for a suitable E_∞ ring spectrum j. We also show that BSF splits as BImJ × BCokerJ as an infinite loop space when localized at an odd prime p, and that, at 2, there is a (non-splittable) infinite loop fibration B Coker J → BSF → B Im J.

Chapter IX contains a theory of pairings in infinite loop space theory. This is used to compare our machine-built spectra of algebraic K-theory to the spectra constructed by Gersten and Wagoner [30,79].

Logically, this book is a sequel to [45 and 46]. However, I have tried to make it self-contained modulo proofs. Thus the definitions of operads and E_∞ spaces are recalled in VI§ 1, and the main results of the cited papers are stated without proof in VII § 1-3. Nevertheless, the reader may find a preliminary reading of the first three sections of [45] helpful, as they contain a leisurely explanation of the motivation behind the basic definitions. While a full understanding of the constructions used in VII requires preliminary reading of [45,§9 and §11], the pragmatic (and trusting) reader may regard the results of that chapter as existence statements derived by means of a black box, the internal intricacies of which can safely be ignored in the applications of the remaining chapters.

It is to be stressed, however, that all of our applications which go beyond the mere assertion that a given space is an infinite loop space depend on special features -- the new multiplicative structure various consistency statements, flexibility in the choice of raw materials -- of our black box which allow us to fit together different parts of the theory.

We illustrate this point with a discussion of how BO appears in our theory. As explained in chapter I, the ordinary classifying space of the infinite orthogonal group is an E_∞ space and thus an infinite loop space. As explained in chapter VI, $\coprod_n BO(n)$ is an E_∞ ring space (and the relevant E_∞ operad is different from that used in chapter I); the zeroth space of the resulting E_∞ ring spectrum is equivalent to $BO \times Z$. We thus have two infinite loop space structures on BO corresponding to two machine-built connective spectra. If we are to take these structures seriously, then we must prove that the machine-built spectra are equivalent to that obtained from the periodic Bott spectrum by killing its homotopy groups in negative degrees. (Other manufacturers of black boxes have not yet studied such consistency problems.) For our first model, the required proof follows from a commutation relation between looping and delooping. For our second model, the required proof follows directly from the ring structure. In both cases, we rely on a characterization of the connective spectrum associated to a periodic space which only makes sense because of special features of our new construction of the stable homotopy category. That both models are necessary can be seen most clearly in the orientation sequence for kO-oriented stable spherical fibrations. This is a fibration sequence of infinite loop spaces

$$SF \to BO_{\otimes} \to B(SF;kO) \to BSF$$

which is derived by use of the E_∞ structure on SF coming from chapter I together with the E_∞ ring structure on kO given by the second model. The first model is essential to relate this sequence to the natural map $j: SO \to SF$ on the infinite loop level. Many of our applications of chapters V and VIII center around this sequence, and its derivation really seems to require every bit of our general abstract machinery. That even the much simpler consistency question about Bott periodicity is not altogether trivial is indicated by the fact that our easier second proof will apply equally well to prove the result about (additive) Brauer lifting cited above.

Beyond its new results, this book is intended to give a coherent account of the most important descriptive (as opposed to computational) applications of infinite loop space theory. A thorough study of the homology of E_∞ spaces and of E_∞ ring space appears in [26], where the theory and results of the present volume are applied to the study of characteristic classes. An informal summary of the material in both that volume and this one is given in [49], which gives an extended intuitive introduction to this general area of topology.

It is a pleasure to acknowledge my debt to the very many people who have helped me with this work. I owe details to Don Anderson, Zig Fiedorowicz, Dick Lashof, Arunas Liulevicius, Stewart Priddy, Vic Snaith, Mark Steinberger, Dick Swan, and Larry Taylor. I am particularly indebted to my coauthors Frank Quinn, Nigel Ray, and Jørgen Tornehave for their ideas and insights, to Ib Madsen for key discussions of 2-primary phenomena and correspondence about various aspects of this work, and to Frank Adams for proofs of a number of technical lemmas and much other help. I owe a special debt to Jim Stasheff for his careful reading of the entire manuscript and his many suggestions for its improvement.

Finally, my thanks to Maija May for preparing the index.

I. $\mathcal{I}$ functors

In [19], Boardman and Vogt introduced the concept of $\mathcal{I}$-functor. Their purpose was to show how certain collections of spaces, such as BFV or BTopV, indexed on inner product spaces V produce E_∞ spaces and therefore infinite loop spaces by passage to limits over $V \subset R^\infty$. In section 1, we give a detailed exposition of this part of their theory, reformulated in terms of operads and operad actions as defined in [45,§1]. In particular, we give a systematic discusssion of the classical groups and their homogeneous spaces as $\mathcal{I}$-functors and display the Bott maps as morphisms of $\mathcal{I}$-functors. In section 2, we relate the two-sided geometric bar construction to $\mathcal{I}$-functors and rederive the theorems of Boardman and Vogt to the effect that F, Top, PL and the related classifying spaces and homogeneous spaces are infinite loop spaces. (We shall use brief ad hoc arguments based on the triangulation theorem to handle PL.) Most of the material of these sections dates from 1971 and has been circulating in preprint form since 1972. Boardman and Vogt's own account of their theory has since appeared [20]. Their language and choice of details are quite different from ours, and there is very little overlap. The present language and results will be needed in the rest of this book.

1. Linear isometries and classical groups

Let $\mathcal{T}$ denote the category of compactly generated and nondegenerately based weak Hausdorff spaces. For an operad $\mathcal{C}$, let $\mathcal{C}[\mathcal{T}]$ denote the category of $\mathcal{C}$-spaces [45, §1] (or VI,§1).

To obtain an action of an operad on the infinite classical groups and related spaces, it is conceptually and notationally simplest to pass from a functor defined on a certain category $\mathcal{J}$ to an action by a related operad $\mathcal{L}$. The definition and properties of $\mathcal{J}$ and $\mathcal{L}$ are due to Boardman and Vogt [19, 20].

Definition 1.1. Define the category $\mathcal{J}$ of linear isometries as follows. The objects of $\mathcal{J}$ are finite or countably infinite dimensional real inner product spaces, topologized as the limits of their finite dimensional subspaces. The morphisms $\mathcal{J}(V, W)$ from V to W are linear isometries $V \to W$, and $\mathcal{J}(V, W)$ is given the (compactly generated) compact-open topology. Note that the direct sum $\oplus: \mathcal{J} \times \mathcal{J} \to \mathcal{J}$ is a continuous functor and is commutative, associative, and unital (with unit $\{0\}$) up to coherent natural isomorphism.

Definition 1.2. Define the linear isometries operad $\mathcal{L}$ by $\mathcal{L}(j) = \mathcal{J}((R^\infty)^j, R^\infty)$, where $(R^\infty)^j$ is the direct sum of j copies of R^∞ with its standard inner product; the requisite data are specified by

(a) $\gamma(f; g_1, \ldots, g_k) = f \circ (g_1 \oplus \ldots \oplus g_k)$, $f \in \mathcal{L}(k)$ and $g_i \in \mathcal{L}(j_i)$.

(b) $1 \in \mathcal{L}(1)$ is the identity map.

(c) $(f\sigma)(y) = f(\sigma y)$ for $f \in \mathcal{L}(j)$, $\sigma \in \Sigma_j$, and $y \in (R^\infty)^j$.

In other words, $\mathcal{L}$ is required to be a sub-operad of the endomorphism operad of R^∞ (where R^∞ has basepoint zero).

It is trivial to verify that Σ_j acts freely on $\mathcal{L}(j)$. The following lemma therefore implies that $\mathcal{L}$ is an E_∞ operad. Recall that isometries need not be isomorphisms.

Lemma 1.3. $\mathcal{J}(V, R^\infty)$ is contractible for all inner product spaces V.

Proof. Let $\{e_i \mid i \geq 1\}$, $\{e_i', e_i'' \mid i \geq 1\}$, $\{f_j\}$, and $\{f_j', f_j''\}$ be orthonormal bases for R^∞, $R^\infty \oplus R^\infty$, V, and $V \oplus V$ respectively. Define $\alpha: R^\infty \to R^\infty$ by $\alpha(e_i) = e_{2i}$ and define $\beta: R^\infty \to R^\infty \oplus R^\infty$ by $\beta(e_{2i-1}) = e_i''$

and $\beta(e_{2i}) = e'_i$. Then β is an isomorphism such that $\beta\alpha = i'$, the injection of the first summand. Define a path $H_1: I \to \mathcal{J}(R^\infty, R^\infty)$ from the identity to α and define a path $H_2: I \to \mathcal{J}(V, V \oplus V)$ from i' to i'' by normalizing the obvious linear paths

$$G_1(t)(e_i) = (1-t)e_i + te_{2i} \quad \text{and} \quad G_2(t)(f_j) = (1-t)f'_j + tf''_j .$$

Fix $\gamma \in \mathcal{J}(V, R^\infty)$ and define $H: I \times \mathcal{J}(V, R^\infty) \to \mathcal{J}(V, R^\infty)$ by

$$H(t,k) = \begin{cases} H_1(2t) \circ k & \text{if } 0 \leq t \leq 1/2 \\ \beta^{-1} \circ (k \oplus \gamma) \circ H_2(2t-1) & \text{if } 1/2 \leq t \leq 1 \end{cases}$$

Then $H(0,k) = k$ and $H(1,k) = \beta^{-1} i'' \gamma$, which is independent of k.

We now define $\mathcal{J}$-functors and a functor from $\mathcal{J}$-functors to $\mathcal{L}$-spaces.

<u>Definition 1.4.</u> An $\mathcal{J}$-functor (T, ω) is a continuous functor $T: \mathcal{J} \to \mathcal{T}$ together with a commutative, associative, and continuous natural transformation $\omega: T \times T \to T \circ \oplus$ (of functors $\mathcal{J} \times \mathcal{J} \to \mathcal{T}$) such that

(a) if $x \in TV$ and if $1 \in T\{0\}$ is the basepoint, then

$$\omega(x, 1) = x \in T(V \oplus \{0\}) = TV$$

(b) if $V = V' \oplus V''$, $\dim V' < \infty$, and if $i: V' \to V$ is the inclusion, then $Ti: TV' \to TV$ (which, by (a), is given by $Ti(x) = \omega(x, 1)$, where 1 is the basepoint of TV'') is a homeomorphism onto a closed subset, and

(c) $TV = \varinjlim TV'$ as a space, where V' runs over the finite dimensional sub inner product spaces of V.

We call ω the Whitney sum; for $x_i \in TV_i$, $1 \leq i \leq j$, we write $\omega(x_1, \ldots, x_j) = x_1 \oplus \ldots \oplus x_j$. A morphism $\Phi: (T, \omega) \to (T', \omega')$ of $\mathcal{J}$-functors is a continuous natural transformation $\Phi: T \to T'$ which commutes with the Whitney sums. $\mathcal{J}[\mathcal{T}]$ denotes the category of $\mathcal{J}$-functors.

Remarks 1.5. (i) The category $\mathcal{J}[\mathcal{T}]$ has finite products; if $t: T' \times T \to T \times T'$ is the interchange natural transformation, then

$$(T, \omega) \times (T', \omega') = (T \times T', (\omega \times \omega')(1 \times t \times 1)) .$$

Similarly, the category $\mathcal{J}[\mathcal{T}]$ has fibred products.

(ii) For $C \in \mathcal{T}$ and an $\mathcal{J}$-functor (T,ω), define the function space $\mathcal{J}$-functor $F(C, T)$ by $F(C, T)(V) = F(C, TV)$, with Whitney sum the composite

$$F(C, TV) \times F(C, TW) \to F(C \times C, TV \times TW) \xrightarrow{F(\Delta, \omega)} F(C, T(V \oplus W)).$$

(iii) If U is the universal covering space functor and if (T, ω) is an $\mathcal{J}$-functor, then $(UT, U\omega)$ is an $\mathcal{J}$-functor, where $U\omega$ is induced from $P\omega$ on $PT = F(I, T)$ by passage to quotient spaces (compare [26, I. 4. 8]).

Definition 1.6. Define a functor $\Theta: \mathcal{J}[\mathcal{T}] \to \mathcal{L}[\mathcal{T}]$ by letting $\Theta(T, \omega) = (TR^{\infty}, \theta)$ on objects and $\Theta(\Phi) = \Phi: TR^{\infty} \to T'R^{\infty}$ on morphisms, where $\theta_j: \mathcal{L}(j) \times (TR^{\infty})^j \to TR^{\infty}$ is defined by

$$\theta_j(f, x_1, \ldots, x_j) = (Tf)(x_1 \oplus \ldots \oplus x_j), \quad f \in \mathcal{L}(j) \text{ and } x_i \in TR^{\infty}.$$

θ_j is continuous by the continuity of T and ω. Observe that Θ commutes with the various constructions specified in the previous remarks, where these constructions are defined on $\mathcal{L}[\mathcal{T}]$ by [45, 1.5-1.7 and 26, I. 4. 8]. As is customary, we shall often write T both for $\mathcal{J}$-functors and for the derived $\mathcal{L}$-spaces TR^{∞}.

Remark 1.7. Let $\alpha: R^{\infty} \to R^{\infty}$ be a linear isometric isomorphism. Then α determines an automorphism $\mathcal{L}\alpha$ of the operad $\mathcal{L}$ by $(\mathcal{L}\alpha)(f) = \alpha f(\alpha^{-1})^j: (R^{\infty})^j \to R^{\infty}$ for $f \in \mathcal{L}(j)$. If (T, ω) is an $\mathcal{J}$-functor, then $T\alpha: TR^{\infty} \to TR^{\infty}$ is an $\mathcal{L}\alpha$-equivariant homeomorphism, in the sense that $T\alpha \circ \theta_j = \theta_j \circ (\mathcal{L}\alpha \times (T\alpha)^j)$.

To construct $\mathcal{L}$-spaces, we need only construct $\mathcal{J}$-functors. We next show that to construct $\mathcal{J}$-functors we need only study finite-dimensional inner product spaces and their linear isometric isomorphisms.

Definition 1.8. Let $\mathcal{J}_n$, $n < \infty$, be the full subcategory of $\mathcal{J}$ whose objects are n-dimensional, and let $\mathcal{J}_*$ be the graded subcategory of $\mathcal{J}$ consisting of the union of the $\mathcal{J}_n$. Note that the functors $\oplus: \mathcal{J}_m \times \mathcal{J}_n \to \mathcal{J}_{m+n}$ together define a graded functor $\oplus: \mathcal{J}_* \times \mathcal{J}_* \to \mathcal{J}_*$. An $\mathcal{J}_*$-functor (T, ω) is a continuous functor $T: \mathcal{J}_* \to \mathcal{T}$ together with a commutative, associative, and continuous natural transformation $\omega: T \times T \to T \circ \oplus$ such that

(a) if $x \in TV$ and if $1 \in T\{0\}$ is the basepoint, then

$$\omega(x, 1) = x \in T(V \oplus \{0\}) = TV .$$

(b) if $V = V' \oplus V''$, $\dim V < \infty$, then the map $TV' \to TV$ given by $x \to \omega(x, 1)$ is a homeomorphism onto a closed subset.

Morphisms of $\mathcal{J}_*$-functors are defined in the evident way, and $\mathcal{J}_*[\mathcal{T}]$ denotes the category of $\mathcal{J}_*$-functors.

Proposition 1.9. The forgetful functor $\mathcal{J}[\mathcal{T}] \to \mathcal{J}_*[\mathcal{T}]$ is an isomorphism of categories.

Proof. We must verify that an $\mathcal{J}_*$-functor (T, ω) admits a unique extension to an $\mathcal{J}$-functor. If $\dim V = \infty$, we can and must define TV by Definition 1.4(c); we shall write $x' \oplus 1$ for the image in TV of $x' \in TV'$ (since $V = V' \oplus (V')^{\perp}$ when $\dim V' < \infty$). Similarly, we can and must define $\omega: TV \times TW \to T(V \oplus W)$ by $x \oplus y = (x' \oplus y') \oplus 1$ if $x = x' \oplus 1$ and $y = y' \oplus 1$ with $x' \in TV'$ and $y' \in TW'$ for finite dimensional subspaces V' of V and W' of W. Finally, if $f: V \to W$ is a linear isometry and if $x = x' \oplus 1 \in TV$, with $x' \in TV'$ where $\dim V' < \infty$, we define

(1) $\quad (Tf)(x) = (Tf')(x') \oplus 1 \quad , \quad$ where $f' = f|V': V' \to f(V')$.

This definition is forced since the image of $f'' = f|(V')^{\perp}$ is contained in $f(V')^{\perp}$, hence $f = f' \oplus f''$, and we therefore must have

$$(Tf)(x) = T(f' \oplus f'')(x' \oplus 1) = (Tf')(x') \oplus (Tf'')(1) = (Tf')(x') \oplus 1,$$

the last equality holding since Tf'' must preserve basepoints. It is straightforward to verify that (T, ω), so constructed, is indeed a well-defined $\mathcal{J}$-functor. Similarly, morphisms of $\mathcal{J}_*$-functors extend uniquely to morphisms of $\mathcal{J}$-functors by passage to limits.

Henceforward, we shall identify the categories $\mathcal{J}_*[\mathcal{T}]$ and $\mathcal{J}[\mathcal{T}]$. We shall speak of $\mathcal{J}$-functors but shall only construct the underlying $\mathcal{J}_*$-functors. The following remarks will be basic to the applications.

Remarks 1.10. For many of the $\mathcal{J}$-functors (T, ω) of interest, the points $x \in TV$ will be (or will be derived from) maps $tV \to tV$, for some space tV depending functorially on V; the basepoint of TV will be the identity map of tV. Moreover, when $\dim V = \dim W < \infty$, a point $f \in \mathcal{J}(V, W)$ will determine a homeomorphism $tf: tV \to tW$ and we will have

$$(2) \qquad (Tf)(x) = tf \circ x \circ (tf)^{-1} \quad \text{for} \quad x \in TV.$$

Henceforward, we shall generally replace formulas (1) and (2) by the notationally simpler expression

$$(Tf)(x) = fxf^{-1} \quad \text{for } x \in TV \text{ and } f \in \mathcal{J}(V, W).$$

We thus suppress from the notation both the passage from f to tf and the required restriction to finite dimensional subspaces. It will often be the case that TV is a sub topological monoid of the monoid (under composition) of maps $tV \to tV$. It will follow that the composition product $c: TV \times TV \to TV$ defines a morphism of $\mathcal{J}$-functors. Indeed, the commutativity of the diagrams

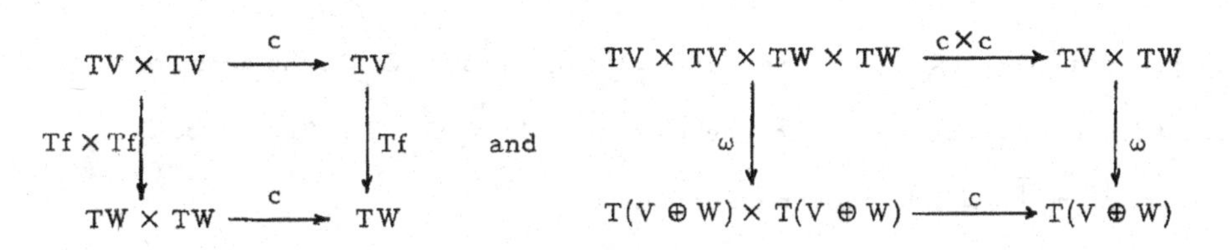

simply amounts to the validity of the formulas

$$fgg'f^{-1} = (fgf^{-1})(fg'f^{-1}) \quad \text{and} \quad gg' \oplus hh' = (g \oplus h)(g' \oplus h')$$

for $f \in \mathcal{J}(V, W)$, $g, g' \in TV$, and $h, h' \in TW$. On TR^{∞}, the composition product will be homotopic to the (internal) Whitney sum induced from the action of $\mathcal{L}$ (see [45, 8.7 or 46, 3.4]). When, further, TV is a sub topological group of the group of homeomorphisms $tV \to tV$, the inverse map $i: TV \to TV$ will define a morphism of $\mathcal{J}$-functors by virtue of the formulas

$$(fgf^{-1})^{-1} = fg^{-1}f^{-1} \quad \text{and} \quad (g \oplus h)^{-1} = g^{-1} \oplus h^{-1}.$$

We now define the classical groups and their homogeneous spaces systematically as $\mathcal{J}$-functors. Let K denote one of the normed division rings R, C, or H (real numbers, complex numbers, or quaternions). For a real inner product space V, let V_K denote $K \otimes_R V$ regarded as a (left) inner product space over K and, for $J \subset K$, let V_K^J denote V_K regarded as an inner product space over J. By a classical group G, we understand any one of the following functors from $\mathcal{J}$ to the category of topological groups:

$$\begin{array}{lllllllll}
e \subset & SO(V) & \subset & O(V) & & & & & \\
& \cap & & \cap & & & & & \\
& SU(V_C) \subset & U(V_C) & \subset & SO(V_C^R) & \subset & O(V_C^R) & & \\
& & \cap & & \cap & & \cap & & \\
& & Sp(V_H) & \subset & SU(V_H^C) & \subset & U(V_H^C) \subset & SO(V_H^R) \subset & O(V_H^R) \quad .
\end{array}$$

(The vertical inclusions are all of the form $g \to 1 \otimes g$.) If G is a classical group and if G_i, $1 \leq i \leq n$, are classical subgroups of G (in the sense of the lattice above), then the homogeneous space $G/\times_{i=1}^{n} G_i$ is defined to be the functor $\mathcal{J} \to \mathcal{T}$ which is given on objects by

$$(G/\times_{i=1}^{n} G_i)(V) = G(V^n)/\times_{i=1}^{n} G_i(V).$$

For $f \in \mathcal{J}(V, W)$, $\dim V = \dim W < \infty$, $(G/\times_{i=1}^{n} G_i)(f)$ is induced by passage to orbit spaces from $g \to f^n g(f^{-1})^n$ for $g \in G(V^n)$, where $f^n: V^n \to W^n$ is the direct sum of f with itself n times. With Whitney sum induced by passage to orbit spaces from the composite

$$G(V^n) \times G(W^n) \xrightarrow{\oplus} G(V^n \oplus W^n) \xrightarrow{G(\nu)} G((V \oplus W)^n),$$

where ν is the evident shuffle isometry, each $(G/\times_{i=1}^{n} G_i, \omega)$ is an $\mathcal{J}$-functor. If $H \subset G$ and $H_i \subset G_i$, then the evident maps of orbit spaces define a morphism of $\mathcal{J}$-functors $H/\times_{i=1}^{n} H_i \to G/\times_{i=1}^{n} G_i$; when $H = G$, the inclusion of the fibre $\times_{i=1}^{n} G_i/H_i$ in $G/\times_{i=1}^{n} H_i$ is also a morphism of $\mathcal{J}$-functors. By the universal bundle of a classical group G, we understand the morphism of $\mathcal{J}$-functors

$$\pi: EG = G/e \times G \to G/G \times G = BG$$

obtained by setting $n = 2$, $G_1 = e$ or G, and $G_2 = G$ in the framework above. Remarks 1.10 show that the product and inverse map on each classical group G are morphisms of $\mathcal{J}$-functors.

We can define Spin by letting Spin(V) be the universal cover of SO(V). Alternatively, and preferably, Spin and also Spin^c, Pin, and Pin^c can be explicitly described as $\mathcal{J}$-functors by means of their standard descriptions in terms of the Clifford algebras of inner product spaces [12]. The product and inverse maps on these groups and the usual maps between these groups and classical groups are then morphisms of $\mathcal{J}$-functors.

We observe next that each of the Bott maps may be regarded as a morphism of $\mathcal{J}$-functors of the form

$$\beta: H/H_1 \times H_2 \rightarrow \Omega(G/G_1 \times G_2).$$

In fact, each β is induced by passage to orbit spaces from a map $\beta: H(V^2) \rightarrow \Omega G(V^2)$ of the form

$$\beta(g)(t) = g\alpha(t)g^{-1}\alpha^{-1}(t), \quad g \in H(V^2) \text{ and } 0 \leq t \leq 1.$$

Here $\alpha(t): V_K^2 \rightarrow V_K^2$ is a linear isometric isomorphism of the general form

$$\alpha(t)(v, v') = (v\alpha_1(t), v'\alpha_2(t)) \quad \text{for } v, v' \in V_K$$

where the $\alpha_i(t)$ are elements of norm one in the relevant ground field K. For example,

$$\beta: BU = U/U \times U \rightarrow \Omega(SU/e\times e) = \Omega SU$$

is so determined by $\alpha_1(t) = e^{\pi i t}$ and $\alpha_2(t) = e^{-\pi i t}$. Explicit definitions of the $\alpha_i(t)$ required in the real case may be found in [21 and 25]. The verification that each β is a morphism of $\mathcal{J}$-functors is an easy calculation from the form of the maps (and explicit expressions for the $\alpha_i(t)$ are not needed). The point is that if $f \in \mathcal{J}(V, W)$, then f_K^2 commutes with $\alpha_i(t)$.

In order to iterate the Bott maps, it is necessary to use the natural maps $\zeta: G \rightarrow \Omega BG$ for $G = O, U$, and Sp. Propositions 2.3 and 2.4 below show that, after passage to $\mathcal{L}$-spaces, these maps become composites of $\mathcal{L}$-maps and of homotopy inverses of $\mathcal{L}$-maps which are homotopy equivalences. As our definition of the stable category and our arguments in II §3 will make clear, these inverses will not complicate the following discussion.

Definition 1.6 yields a natural structure of $\mathcal{L}$-space on the homogeneous spaces $G/\mathop{\times}\limits_{i=1}^{n} G_i = (G/\mathop{\times}\limits_{i=1}^{n} G_i)(R^\infty)$, on Spin, etc. Each of these spaces is grouplike (π_0 is a group) and is thus an infinite loop space by [46, 2.3] (or VII, 3.2 below). Certain of these spaces are also infinite loop spaces by Bott periodicity. To show the consistency of these structures, let X be one of the spaces entering into Bott periodicity of period d, d = 2 or d = 8, such as $\Omega^2 BU \simeq BU \times Z$ or $\Omega^8 BO \simeq BO \times Z$. The $\mathcal{L}$-space structure on X determines a connective coordinatized spectrum (or infinite loop sequence) $B_\infty X = \{B_i X\}$ such that $B_0 X$ is naturally equivalent to X. The iterated Bott map $\beta: X \to \Omega^d X$ induces a weak homotopy equivalence $B_\infty \beta: B_\infty X \to B_\infty \Omega^d X$ of spectra. As will be explained in VII.3.4, [46, 3.1] gives a map $B_\infty \Omega^d X \to \Omega^d B_\infty X$ in the stable category, the zeroth level of which is equivalent to the identity map of $\Omega^d X$. By composition, we therefore have a map $B_\infty X \to \Omega^d B_\infty X$, the zeroth level of which is equivalent to $\beta: X \to \Omega^d X$. It is intuitively obvious, and will be rigorously proven in II §3, that the connective spectrum associated to the periodic spectrum with zeroth space X determined by β is characterized, up to isomorphism in the stable category, by precisely these conditions.

2. The bar construction; F, Top, and PL.

The two-sided geometric bar construction will play a central role in our theory, and the following notations will be used throughout the later chapters. Let G be a topological monoid the identity element of which is a nondegenerate basepoint and let X and Y be left and right G-spaces. Then there is a simplicial topological space $B_*(Y, G, X)$, the n-th space of which is $Y \times G^n \times X$ [45, §10]. Its geometric realization [45, 11.1] will be denoted $B(Y, G, X)$. We shall always write

$$p: B(Y,G,X) \to B(Y,G,*) \quad \text{and} \quad q: B(Y,G,X) \to B(*,G,X)$$

for the maps induced from the trivial G-maps $X \to *$ and $Y \to *$, where $*$ is the one point G-space. p and q are quasi-fibrations with fibres X and Y when G is grouplike and are G-bundles when G is a topological group [47, 7.6 and 8.2]. We shall write

$$\tau = \tau(\rho): Z \to B(Y,G,X) \quad \text{and} \quad \varepsilon = \varepsilon(\lambda): B(Y,G,X) \to Z$$

for the maps induced by a map $\rho: Z \to Y \times X$ and by a map $\lambda: Y \times X \to Z$

such that $\lambda(yg, x) = \lambda(y, gx)$ [45,9.2]; the intended choice of ρ and λ should be clear from the context. By [45, 9.8, 9.9, and 11.10], the G-maps

$$\varepsilon : B(Y, G, G) \to Y \quad \text{and} \quad \varepsilon : B(G, G, X) \to X$$

are deformation retractions with right inverses the evident maps τ. We agree to abbreviate $B(\alpha, \gamma, \beta) = B\alpha$ if γ and β are identity maps, and similarly in the other variables.

$BG = B(*, G, *)$ is the standard classifying space of G. If G is group-like, the natural inclusion $\zeta: G \to \Omega BG$ is a weak homotopy equivalence. If, further, $p': E' \to B'$ is a principal quasi G-fibration with E' aspherical, then the maps

$$B' \xleftarrow{\varepsilon(p')} B(E', G, *) \xrightarrow{q} BG$$

display a weak homotopy equivalence between B' and BG [47, 8.7 and 7.7].

For a morphism $j: H \to G$ of grouplike topological monoids, define the left and right "homogeneous spaces" by

$$G/H = B(G, H, *) \quad \text{and} \quad H\backslash G = B(*, H, G).$$

There is a weak homotopy equivalence $\theta: G/H \to FBj$, where FBj denotes the homotopy theoretic fibre, and a quasifibration sequence

$$H \xrightarrow{j} G \xrightarrow{\tau} G/H \xrightarrow{q} BH \xrightarrow{Bj} BG$$

[47, 8.8]. By symmetry, the same statements hold for $H\backslash G$.

Now let $\mathcal{C}$ be any operad. As explained in [46,§3], $B(Y, G, X)$ is a $\mathcal{C}$-space when Y, G, and X are $\mathcal{C}$-spaces and the product and unit of G and its action on Y and on X are morphisms of $\mathcal{C}$-spaces. Such a G is said to be a monoid in $\mathcal{C}[\mathcal{J}]$; its monoid product is homotopic to the product given by the action of $\mathcal{C}$ [46,3.4]. When all given maps in any of the constructions of the previous paragraphs are morphisms of $\mathcal{C}$-spaces, then so are all derived maps (where homotopies $X \times I \to Y$ are interpreted as maps $X \to F(I^+, Y)$ and where the fibre of a $\mathcal{C}$-map is a $\mathcal{C}$-space via

[45, 1.8]). For most of the maps above, this statement follows from the fact that geometric realization defines a functor from simplicial $\mathcal{C}$-spaces to $\mathcal{C}$-spaces [45, 12.2]. For the remaining maps, easy direct calculations as in [46, 3.6], where $\zeta: G \to \Omega BG$ is handled, are required.

We next show that precisely similar statements apply to $\mathcal{I}$-functors.

Definition 2.1. A monoid-valued $\mathcal{I}$-functor, or monoid in $\mathcal{I}[\mathcal{T}]$, is an $\mathcal{I}$-functor G such that each GV is a topological monoid, the identity element of GV is its basepoint, and the products $GV \times GV \to GV$ define a morphism of $\mathcal{I}$-functors. G is said to be group-valued if each GV is a topological group and the inverse maps $GV \to GV$ define a morphism of $\mathcal{I}$-functors. G is said to be grouplike if each $\pi_0 GV$ is a group. A left action of a monoid-valued $\mathcal{I}$-functor G on an $\mathcal{I}$-functor X is a morphism of $\mathcal{I}$-functors $G \times X \to X$ such that the map $GV \times XV \to XV$ is an action of GV on XV for each V.

Definition 2.2. Let G be a monoid-valued $\mathcal{I}$-functor which acts from the left and right on $\mathcal{I}$-functors X and Y. Define an $\mathcal{I}$-functor $B(Y, G, X)$ as follows. For $V \in \mathcal{I}$ and for a morphism $f: V \to V'$ in $\mathcal{I}$, define

$$B(Y, G, X)(V) = B(YV, GV, XV) \quad \text{and} \quad B(Y, G, X)(f) = B(Yf, Gf, Xf).$$

The Whitney sum on $B(Y, G, X)$ is defined by the composite maps

$$\begin{array}{c} B(YV, GV, XV) \times B(YW, GW, XW), \\ \downarrow \xi \\ B(YV \times YW,\ GV \times GW,\ XV \times XW) \\ \downarrow B(\omega, \omega, \omega) \\ B(Y(V \oplus W),\ G(V \oplus W),\ X(V \oplus W)) \end{array}$$

where ξ is the commutative and associative natural homeomorphism given by [45, 10.1 and 11.5].

It is easily verified that B(Y, G, X) is well-defined. The only point worth mentioning is that the functor B(Y, G, X) is continuous because geometric realization is continuous if the set of maps of simplicial spaces from C to D is topologized as a subspace of the product over n of the spaces of maps from C_n to D_n. Clearly the $\mathcal{L}$-space obtained by first applying the functor B to (Y, G, X) and then applying the functor $\oplus$ of Definition 1.6 coincides with the $\mathcal{L}$-space obtained by first applying $\oplus$ to Y, G, and X and then applying B.

In view of Remarks 1.5, all of the material of the first few paragraphs of this section can be rephrased in terms of $\mathcal{J}$-functors (where a homotopy between maps $T \to T'$ of $\mathcal{J}$-functors means a morphism of $\mathcal{J}$-functors $T \to F(I^+, T')$ and where, just as for spaces, the fibre of a map of $\mathcal{J}$-functors is defined as the evident fibred product). It is again straightforward to verify that all constructed maps are morphisms of $\mathcal{J}$-functors when all given maps are so. We summarize results in the following statements.

Proposition 2.3. If $j: H \to G$ is a morphism of grouplike monoid-valued $\mathcal{J}$-functors, then the following is a quasi-fibration sequence of $\mathcal{J}$-functors

$$H \xrightarrow{j} G \xrightarrow{\tau} G/H \xrightarrow{q} BH \xrightarrow{Bj} BG .$$

$\zeta: G \to \Omega BG$ and $\theta: G/H \to FBj$ are weak equivalences of $\mathcal{J}$-functors.

The classical groups define group-valued $\mathcal{J}$-functors by Remarks 1.10.

Proposition 2.4. When G is a classical group $\mathcal{J}$-functor, the maps

$$G/G \times G \xleftarrow{\varepsilon(\pi)} B(G/e \times G, G, *) \xrightarrow{q} BG$$

of $\mathcal{J}$-functors determine a weak equivalence of $\mathcal{L}$-spaces between the two natural classifying spaces of G.

Henceforward, we shall work relative to a fixed continuous sphere-valued functor $t: \mathcal{J}_* \to \mathcal{T}$ such that $tV \wedge tW = t(V \oplus W)$. By the continuity, commutativity, and associativity of the smash product, t is itself an $\mathcal{J}_*$-functor with the natural projection $tV \times tW \to t(V \oplus W)$ as Whitney sum. We have the following list of monoid-valued $\mathcal{J}_*$-functors which act from the left on t.

<u>Examples 2.5.</u> Let $\widetilde{F}$ denote the $\mathcal{J}_*$-functor specified by

$$\widetilde{F}V = F(tV, tV) \quad \text{and} \quad \widetilde{F}f = F(tf^{-1}, tf) \quad \text{for} \quad f: V \to V',$$

with the Whitney sum $\omega: \widetilde{F}V \times \widetilde{F}W \to \widetilde{F}(V \oplus W)$ given by the smash product of maps. Define sub $\mathcal{J}_*$-functors F, SF, Top, and STop of $\widetilde{F}$ by restricting attention to based homotopy equivalences, degree one homotopy equivalences, homeomorphisms, and degree one homeomorphisms of tV. Define e to be the trivial sub $\mathcal{J}_*$-functor of $\widetilde{F}$, $eV = \{1\}$. When tV is the one-point compactification V^* of V, O and SO become sub $\mathcal{J}_*$-functors of Top and STop via one-point compactification of maps. Moreover, the twisted adjoint representation of [12, p. 7] defines morphisms of $\mathcal{J}_*$-functors from Pin, Spin, Γ, Pin^c, Spin^c, and Γ^c to O. When $tV = V_C^*$, U and SU become sub $\mathcal{J}_*$-functors of STop via one-point compactification, and similarly for Sp when $tV = V_H^*$. By Remarks 1.10, and [12], all of these $\mathcal{J}_*$-functors are monoid-valued and all but SF, F, and $\widetilde{F}$ are group-valued.

We pass from $\mathcal{J}_*$-functors to $\mathcal{J}$-functors by Proposition 1.9. Write $\widetilde{F}R^n = \widetilde{F}(n)$, $\widetilde{F}R^\infty = \widetilde{F}$, and similarly for other $\mathcal{J}$-functors. It is usual to define G(n) to be the space of all homotopy equivalences of S^{n-1} and to regard F(n-1) as the subspace of based homotopy equivalences. It is clear that $G = \varinjlim G(n)$ is then homotopy equivalent to F. G is the more natural space to consider in some geometric situations, but we shall work with F since it is this space which occurs naturally in our theory.

We pass from $\mathcal{J}$-functors to $\mathcal{L}$-spaces via Definition 1.6 and from $\mathcal{L}$-spaces to infinite loop spaces via [45 and 46] (or VII §3). Thus F, SF, Top, STop, the classical groups, Spin, $Spin^c$, and related groups and all of their classifying spaces and homogeneous spaces are infinite loop spaces, and all of the natural maps between these spaces are infinite loop maps.

It remains to consider PL and related spaces. One could perhaps develop a complete geometric theory by introducing categories with the same objects as $\mathcal{J}$ but with simplicial sets of morphisms from V to W, the k-simplices of which would be appropriate piecewise linear and piecewise differential fibrewise homeomorphisms $\Delta_k \times V \to \Delta_k \times W$; here PD would be required in order to relate PL to O. I have not attempted to go through the details.

A second approach would be to consider the $\mathcal{J}$-functor "PL" such that "PLV" is the subspace of FV consisting of the based piecewise linear homeomorphisms of V^*. Unfortunately, "PL" has the wrong homotopy type; according to Rourke (private communication), the inclusion of "PL" in Top is a weak homotopy equivalence. It is at least conceivable that a larger topology on "PL" exists which does have the right homotopy type.

Our approach is to ignore these difficulties, to recall that PL was largely introduced in order to study the triangulation problem, and to observe that the homotopy types of PL, BPL, PL/O, etc. are completely determined as infinite loop spaces by the solution to this problem and by the infinite loop space structures already derived on Top, BTop, Top/O, etc.

In detail, we note that $Top/PL = K(Z_2, 3)$, Top/O is a 2-connected space, $\pi_3(Top/O) = Z_2$, and the natural map $Top/O \to Top/PL$ induces an isomorphism on π_3. Recall that there is only one (n-1)-connected spectrum

$\mathcal{K}(Z_2, n)$ with zeroth space a $K(Z_2, n)$ (up to isomorphism in any good stable category) and there is only one non-trivial map from an (n-1)-connected spectrum with $\pi_n E = Z_2$ into $\mathcal{K}(Z_2, n)$. Thus Top/PL, however it is constructed geometrically, is just $K(Z_2, 3)$ as an infinite loop space, and the unique non-trivial map Top/O $\rightarrow$ Top/PL is an infinite loop map in precisely one way.

Now define the following spaces (or rather homotopy types, since that is all our data determine). In each case, it is clear that any permissible geometric construction of the space named must yield the specified homotopy type.

(a) PL/O is the fibre of the unique non-trivial map Top/O $\rightarrow K(Z_2, 3)$.

(b) PL is the fibre of the composite Top $\rightarrow$ Top/O $\rightarrow K(Z_2, 3)$ and SPL is the fibre of the restriction of this composite to STop.

(c) BPL is the fibre of the composite BTop $\rightarrow B_1$(Top/O) $\rightarrow K(Z_2, 3)$ and BSPL is the fibre of the restriction of this composite to BSTop. (Here BTop is equivalent to the delooping B_1Top by VII. 3. 5.)

(d) F/PL is the fibre of the composite BPL $\rightarrow$ BTop $\rightarrow$ BF.

Clearly the fibre of an infinite loop map is an infinite loop space, and it follows that each space we have constructed has a well-defined infinite loop space structure such that all of the natural maps between these spaces are infinite loop maps. Similar constructions handle PL/G and related maps for other classical groups G.

II. Coordinate-free spectra

A spectrum T is usually defined to be a sequence of spaces T_i and maps $\sigma_i: \Sigma T_i \to T_{i+1}$. Let $\{e_n\}$ be the standard basis for R^∞ and think of the 1-sphere as the one-point compactification tRe_n of the subspace Re_n of R^∞. Then a change of notation allows us to describe T as a sequence of spaces $T(R^i)$ and maps $\sigma_i: T(R^i) \wedge tRe_{i+1} \to T(R^{i+1})$. Thus the usual notion of spectrum implicitly refers to a fixed chosen basis for R^∞. Many very real difficulties in the homotopy theory of spectra, in particular the problems associated with the construction of well-behaved smash products, arise from permutations of suspension coordinates. Such permutations can be thought of as resulting from changes of basis for R^∞, and we shall see in [48] that the coordinate-free definition of spectra to be given here leads to a relatively simple development of the properties of the stable homotopy category.

However, our present concern is with more than just stable homotopy theory. In order to define E_∞ ring spectra, it is essential to work in a category of (omega) spectra which enjoys good properties even before passage to homotopy. The point is that these spectra have very rich internal structure, much of which is lost upon passage to the homotopy category.

The spectra used in the best previous constructions of stable homotopy categories are (or are derived from) CW-spectra, namely those spectra T such that T_i is a CW-complex and each σ_i is a cellular inclusion. Obviously such rigid structures cannot possibly be related to infinite loop spaces before passage to homotopy, and our spectra will be cell-free as well as coordinate-free. Restriction to CW-spectra is in any case undesirable since CW-spectra seldom occur in nature and are not closed under such simple and useful constructions as formation of product and loop spectra. In our stable category, desuspension will be given by the loop spectrum functor.

We define coordinate-free prespectra and spectra, show how to pass back and forth between spaces, prespectra, and spectra, and relate coordinate-free to coordinatized spectra in section 1. We define the stable homotopy category and discuss ring spectra, connective spectra, and localizations and completions of spectra in section 2. We shall omit most proofs in these sections; the missing details may be found in [48] and are largely irrelevant to our later work in this book. In section 3, we consider cohomology theories and give a rather pedantic analysis of the precise relationships between periodic spaces, periodic spectra, and "periodic connective spectra".

Although exploited in a wholly different way, the idea of using linear isometries to study the stable category is due to Boardman [18]. Puppe [56] independently came to the idea of coordinate-free prespectra.

1. Spaces, prespectra, and spectra

Recall the definition of $\mathcal{J}_*$ from I.1.8 and, as in I§2, fix a continuous sphere-valued functor $t: \mathcal{J}_* \to \mathcal{T}$ such that $t\{0\} = S^0$ and $tV \wedge tW = t(V \oplus W)$. In practice, tV will be the one-point compactification of τV for some functor $\tau : \mathcal{J} \to \mathcal{J}$, and we shall see in Remarks 1.9 that restriction to the identity functor τ would result in no real loss of generality. Recall that $F(X, Y)$ denotes the space of based maps $X \to Y$ and define the suspension and loop functors based on V to be

$$\Sigma^V X = X \wedge tV \quad \text{and} \quad \Omega^V X = F(tV, X).$$

It will be important to keep in mind the old-fashioned distinction between external and internal direct sums; we write $\oplus$ for the former and $+$ for the latter. We write $V \perp W$ to indicate that two sub inner product spaces of a given inner product space are orthogonal; the notation $+$ will only be used between orthogonal subspaces and will thus carry orthogonality as an implied hypothesis.

Let $\mathcal{J}_*(R^\infty)$ denote the full subcategory of $\mathcal{J}_*$ the objects of which are the finite-dimensional sub inner product spaces of R^∞ and let $\mathcal{H}\mathcal{T}$ denote the category of based spaces and based homeomorphisms. Let $h\mathcal{A}$ denote the homotopy category associated to a topological category $\mathcal{A}$. Since $\mathcal{J}_n(V, W)$ is homeomorphic to $O(n)$, $(h\mathcal{J}_n)(V, W)$ has precisely two elements if $n > 0$. Remarks 1.10 below indicate a possible simplification of the of the following definition.

<u>Definition 1.1.</u> A prespectrum (T, σ) is a function $T: \mathcal{J}_*(R^\infty) \to \mathcal{H}\mathcal{T}$ (on objects and morphisms) which induces a functor $T: h\mathcal{J}_*(R^\infty) \to h\mathcal{T}$ together with based maps $\sigma: \Sigma^W TV \to T(V + W)$ for $V \perp W$ which satisfy the following conditions.

(i) Each adjoint map $\tilde{\sigma}: TV \to \Omega^W T(V + W)$ is an inclusion with closed image.

(ii) The following diagrams commute in $\mathcal{T}$, where $V \perp W \perp Z \perp V$:

$$\begin{array}{ccc} TV & = & \Sigma^0 TV \\ {\scriptstyle 1}\downarrow & & \downarrow{\scriptstyle \sigma} \\ TV & = & T(V + \{0\}) \end{array} \qquad \text{and} \qquad \begin{array}{ccc} \Sigma^Z\Sigma^W TV & = & \Sigma^{W+Z} TV \\ {\scriptstyle \Sigma^Z\sigma}\downarrow & & \downarrow{\scriptstyle \sigma} \\ \Sigma^Z T(V + W) & \xrightarrow{\sigma} & T(V + W + Z) \end{array}$$

(iii) The following diagrams commute in $h\mathcal{T}$, where $f \in \mathcal{J}_*(V, V')$, $g \in \mathcal{J}_*(W, W')$, $V \perp W$ and $V' \perp W'$:

$$\begin{array}{ccc} TV \wedge tW = \Sigma^{W} TV & \xrightarrow{\sigma} & T(V+W) \\ \downarrow{\scriptstyle Tf \wedge tg} & & \downarrow{\scriptstyle T(f+g)} \\ TV' \wedge tW' = \Sigma^{W'} TV' & \xrightarrow{\sigma} & T(V'+W') \end{array}$$

(T, σ) is said to be a spectrum if each $\tilde{\sigma}: TV \to \Omega^{W} T(V+W)$ is a homeomorphism. A morphism $\theta: (T, \sigma) \to (T', \sigma')$ of prespectra consists of maps $\theta: TV \to T'V$ such that $\theta: T \to T'$ is a natural transformation of functors $h\mathcal{I}_*(R^{\infty}) \to h\mathcal{J}$ and such that the following diagrams commute in $\mathcal{J}$ for $V \perp W$:

$$\begin{array}{ccc} \Sigma^{W} TV & \xrightarrow{\sigma} & T(V+W) \\ \downarrow{\scriptstyle \Sigma^{W}\theta} & & \downarrow{\scriptstyle \theta} \\ \Sigma^{W} T'V & \xrightarrow{\sigma} & T'(V+W) \end{array}$$

Let $\mathcal{P}$ denote the category of prespectra and let $\mathcal{S}$ denote its full subcategory of spectra. Let $\nu: \mathcal{S} \to \mathcal{P}$ denote the inclusion functor.

The category $\mathcal{S}$ is of primary interest, and $\mathcal{P}$ is to be regarded as a convenient auxiliary category. The pair of terms (prespectrum, spectrum), which was introduced by Kan, is distinctly preferable to the older pair (spectrum, Ω-spectrum) since spectra are the fundamental objects of study and since prespectra naturally give rise to spectra. The use of homeomorphisms, rather than homotopy equivalences, in the definition of spectra is both essential to the theory and convenient in the applications. We have little use for the classical notion of an Ω-spectrum.

Granted the desirability of a coordinate-free theory of spectra, it is clearly sensible to think of the finite-dimensional subspaces of R^{∞} as an indexing set. Thus a prespectrum ought, at least, to consist of spaces TV and maps $\sigma: \Sigma^{W} TV \to T(V+W)$, and it is obviously reasonable to insist that TV be homeomorphic to TW if $\dim V = \dim W$. Our definition merely codifies these specifications in a coherent way. The use of the homotopy category $h\mathcal{J}_*(R^\infty)$ can be thought of as a systematic device for letting linear isometries keep track of signs, changes of coordinates, and such like complications in the usual theory of spectra. One might be tempted to define prespectra by requiring T to be a functor $\mathcal{J}_*(R^\infty) \to \mathcal{HJ}$ and σ to be natural, without use of homotopy categories, but such a definition would not allow the construction of spectra from prespectra or of coordinate-free spectra from coordinatized spectra.

We next make precise the categorical interrelationships among $\mathcal{J}$, $\mathcal{P}$, and $\mathcal{S}$ (as was done in [43] for ordinary spectra). Observe that we have forgetful, or zeroth space, functors $\mathcal{P} \to \mathcal{J}$ and $\mathcal{S} \to \mathcal{J}$ defined on objects by $(T, \sigma) \to T_0 = T\{0\}$.

Definition 1.2. Define the suspension prespectrum functor $\Sigma^\infty : \mathcal{J} \to \mathcal{P}$ by letting

$$(\Sigma^\infty X)(V) = \Sigma^V X \quad \text{and} \quad (\Sigma^\infty X)(f) = \Sigma^f X \equiv 1 \wedge tf$$

on objects V and morphisms f of $\mathcal{J}_*(R^\infty)$ and by letting

$$\sigma = 1 : \Sigma^W \Sigma^V X \to \Sigma^{V+W} X \quad \text{for } V \perp W.$$

For $\phi : X \to X'$, let $\Sigma^\infty \phi = \Sigma^V \phi : \Sigma^V X \to \Sigma^V X'$.

Lemma 1.3. $\Sigma^\infty X$ is the free prespectrum generated by the space X; that is, for $T \in \mathcal{P}$ and $\phi : X \to T_0$, there is a unique map $\tilde{\phi} : \Sigma^\infty X \to T$ of prespectra with zeroth map ϕ.

Definition 1.4. Define the associated spectrum functor $\Omega^\infty : \mathcal{P} \to \mathcal{S}$ as follows. Let $(T, \sigma) \in \mathcal{P}$. We have identifications

$$(*) \qquad \Omega^{W+Z} T(V + W + Z) = \Omega^W \Omega^Z T(V + W + Z)$$

for $V \perp W \perp Z \perp V$ and we define

$$(\Omega^{\infty}T)(V) = \varinjlim_{W \perp V} \Omega^{W} T(V + W)$$

where the limit is taken with respect to the inclusions

$$\Omega^{W}\tilde{\sigma}, \quad \tilde{\sigma}: T(V + W) \rightarrow \Omega^{Z} T(V + W + Z).$$

For $V \perp W$, the required homeomorphisms

$$\tilde{\sigma}: (\Omega^{\infty}T)(V) \rightarrow \Omega^{W}(\Omega^{\infty}T)(V + W)$$

are obtained by passage to limits over Z from the identifications (*). To define $(\Omega^{\infty}T)(f): (\Omega^{\infty}T)(V) \rightarrow (\Omega^{\infty}T)(V')$ for $f: V \rightarrow V'$, choose Z which contains both V and V', let W and W' be the orthogonal complements of V and V' in Z, and observe that there is one and, up to homotopy, only one linear isometry $g: W \rightarrow W'$ such that $f + g$ is homotopic (through isometries) to the identity of Z. The required homeomorphism $(\Omega^{\infty}T)(f)$ is obtained by passage to limits over $Z' \perp Z$ from the maps

$$\Omega^{W+Z'} T(Z + Z') \rightarrow \Omega^{W'+Z'} T(Z + Z')$$

given by composition with $t(g^{-1} + 1)$. For a map $\theta: T \rightarrow T'$ of prespectra, define $\Omega^{\infty}\theta: (\Omega^{\infty}T)(V) \rightarrow (\Omega^{\infty}T')(V)$ by passage to limits from the maps

$$\Omega^{W}\theta: \Omega^{W}T(V + W) \rightarrow \Omega^{W}T'(V + W).$$

Lemma 1.5. The inclusions $TV = \Omega^{0}T(V + \{0\}) \rightarrow \varinjlim \Omega^{W}T(V + W)$ define a map $\iota: T \rightarrow \nu\Omega^{\infty}T$ of prespectra, and $\Omega^{\infty}T$ is the free spectrum generated by the prespectrum T; that is, for $E \in \mathcal{S}$ and $\theta: T \rightarrow \nu E$, there is a unique map $\tilde{\theta}: \Omega^{\infty}T \rightarrow E$ of spectra such that $\tilde{\theta}\iota = \theta$.

Definition 1.6. Define $Q_{\infty} = \Omega^{\infty}\Sigma^{\infty}: \mathcal{J} \rightarrow \mathcal{S}$ and define $QX = (Q_{\infty}X)_0$. Observe that QX is then homeomorphic to $\varinjlim \Omega^{n}\Sigma^{n}X$. $Q_{\infty}X$ is the free spectrum generated by X, and we thus have an adjunction

$$\mathcal{J}(X, E_0) \cong \mathcal{S}(Q_\infty X, E), \quad X \in \mathcal{J} \quad \text{and } E \in \mathcal{S}.$$

Finally, we define coordinatized prespectra and spectra and relate them to the coordinate-free variety.

Definition 1.7. Let $A = \{A_i\}$, $\{0\} = A_0 \subset A_1 \subset \ldots \subset A_i \subset \ldots$, be an increasing sequence of subspaces of R^∞, with $R^\infty = \bigcup_{i \geq 0} A_i$. Let B_i denote the orthogonal complement of A_i in A_{i+1}. A prespectrum $\{T_i, \sigma_i\}$ indexed by A is a sequence of based spaces T_i and maps $\sigma_i : \Sigma^{b_i} T_i = T_i \wedge tB_i \to T_{i+1}$ such that the adjoints $\tilde{\sigma}_i : T_i \to \Omega^{b_i} T_{i+1}$ are inclusions with closed images. $\{T_i, \sigma_i\}$ is a spectrum if each σ_i is a homeomorphism. A map $\{\theta_i\} : \{T_i, \sigma_i\} \to \{T'_i, \sigma'_i\}$ of prespectra is a sequence of based maps $\theta_i : T_i \to T'_i$ such that $\theta_{i+1} \circ \sigma_i = \sigma'_i \circ \Sigma^{b_i} \theta_i$. Let $\mathcal{P}A$ and $\mathcal{S}A$ denote the category of prespectra indexed by A and its full subcategory of spectra. Define a forgetful functor $\emptyset : \mathcal{P} \to \mathcal{P}A$ by letting $\emptyset(T, \sigma) = \{T_i, \sigma_i\}$, where $T_i = TA_i$ and $\sigma_i = \sigma : \Sigma^{b_i} TA_i \to TA_{i+1}$.

With trivial modifications, the previous definitions and lemmas apply to coordinatized prespectra and spectra. The following result would be false on the prespectrum level.

Theorem 1.8. The forgetful functor $\emptyset : \mathcal{S} \to \mathcal{S}A$ is an equivalence of categories; that is, there is a functor $\psi : \mathcal{S}A \to \mathcal{S}$ and there are natural isomorphisms

$$\emptyset\psi\{E_i, \sigma_i\} \cong \{E_i, \sigma_i\} \quad \text{and} \quad \psi\emptyset(E, \sigma) \cong (E, \sigma).$$

Remark 1.9. Suppose that $tV = \tau V \cup \infty$ for some nontrivial additive functor $\tau : \mathcal{J} \to \mathcal{J}$. Then τR^∞ has countably infinite dimension and we may choose an isometric isomorphism $g : \tau R^\infty \to R^\infty$. Visibly g determines an isomorphism between the category $\mathcal{S}A$ defined by use of t and the category $\mathcal{S}gA$ defined by use of the one-point compactification of inner product spaces. Therefore, up to equivalence, the category $\mathcal{S}$ is independent of the choice of t (of the specified form).

Remark 1.10. Theorem 1.8 implies that explicit use of linear isometries in our definition of a spectrum is quite unnecessary, and the details in Definition 1.4 show why this is the case. I find the introduction of isometries conceptually helpful, particularly on the prespectrum level (compare IV.1.3), but the reader is free to ignore them throughout.

2. The stable homotopy category

We require small smash products and function spectra, "small" meaning between spaces and spectra rather than between spectra and spectra.

Definition 2.1. For $X \in \mathcal{T}$ and $T \in \mathcal{P}$, define $T \wedge X \in \mathcal{P}$ by letting

$$(T \wedge X)(V) = TV \wedge X \quad \text{and} \quad (T \wedge X)(f) = Tf \wedge 1$$

on objects V and morphisms f of $\mathcal{J}_*(R^\infty)$ and by letting the structural maps σ be the composites

$$\Sigma^W(TV \wedge X) \cong (\Sigma^W TV) \wedge X \xrightarrow{\sigma \wedge 1} T(V+W) \wedge X .$$

For $E \in \mathcal{S}$, define $E \wedge X \in \mathcal{S}$ by $E \wedge X = \Omega^\infty(\nu E \wedge X)$.

Observe that Ω^∞ and ν can be used similarly to transport to spectra any functor on prespectra which does not preserve spectra. The functors Σ^∞ and Ω^∞ preserve smash products with spaces.

Lemma 2.2. For $X \in \mathcal{T}$ and $Y \in \mathcal{T}$, $\Sigma^\infty(Y \wedge X)$ is isomorphic to $(\Sigma^\infty Y) \wedge X$. For $X \in \mathcal{T}$ and $T \in \mathcal{P}$, $\Omega^\infty(\iota \wedge 1): \Omega^\infty(T \wedge X) \to (\Omega^\infty T) \wedge X$ is an isomorphism.

Definition 2.3. For $X \in \mathcal{T}$ and $T \in \mathcal{P}$, define $F(X, T) \in \mathcal{P}$ by letting

$$F(X,T)(V) = F(X,TV) \text{ and } F(X,T)(f) = F(1,Tf)$$

on objects V and morphisms f of $\mathcal{J}_*(R^\infty)$ and by letting the structural maps σ be the adjoints of the composites

$$F(X,TV) \xrightarrow{F(1,\tilde{\sigma})} F(X,\Omega^W T(V+W)) \cong \Omega^W F(X,T(V+W)).$$

If T is a spectrum, then so is $F(X,T)$.

Lemma 2.4. For $X \in \mathcal{J}$ and $E \in \mathcal{S}$, $\nu F(X,E) = F(X,\nu E)$. For $X \in \mathcal{J}$ and $T \in \mathcal{P}$, $F(X,T)_0 = F(X,T_0)$.

Lemma 2.5. For $X \in \mathcal{J}$, there are natural (adjunction) isomorphisms

$$\mathcal{P}(T \wedge X, T') \cong \mathcal{P}(T, F(X,T')), \quad T,T' \in \mathcal{P}$$

and

$$\mathcal{S}(E \wedge X, E') \cong \mathcal{S}(E, F(X,E')), \quad E,E' \in \mathcal{S}.$$

Explicitly, the adjoint $\tilde{\theta}: E \to F(X,E')$ of $\theta: E \wedge X \to E'$ has V^{th} map $EV \to F(X,E'V)$ the adjoint in $\mathcal{J}$ of the composite

$$EV \wedge X = (\nu E \wedge X)(V) \xrightarrow{\iota} \nu\Omega^\infty(\nu E \wedge X)(V) = (E \wedge X)(V) \xrightarrow{\theta} E'V.$$

Let K^+ denote the union of a space K and a disjoint basepoint.

Definition 2.6. For Y and Z both in $\mathcal{J}$ or $\mathcal{P}$ or $\mathcal{S}$, define a homotopy $h: f_0 \simeq f_1$ between maps $f_i: Y \to Z$ to be a map $h: Y \wedge I^+ \to Z$ (in the relevant category) such that $h|Y \wedge \{i\}^+ = f_i$. Note that h could equally well be considered as a map $Y \to F(I^+, Z)$. Let $\pi(Y,Z)$ denote the set of homotopy classes of maps $Y \to Z$.

The basic machinery of elementary homotopy theory, such as the dual Barratt-Puppe sequences and dual Milnor $\varprojlim^1$ exact sequences, applies equally well in $\mathcal{J}$ and in $\mathcal{S}$ [48, I]. Lemmas 2.2 and 2.4 imply that Σ^∞, Ω^∞, Q_∞, ν, and the zeroth space functors $\mathcal{P} \to \mathcal{J}$ and $\mathcal{S} \to \mathcal{J}$ are all homotopy preserving. Clearly Σ^∞, Ω^∞, and Q_∞ are still free functors after passage to homotopy categories. In particular, we have a natural isomorphism

$$\pi(X, E_0) \cong \pi(Q_\infty X, E), \quad X \in \mathcal{J} \quad \text{and } E \in \mathcal{S}.$$

Definitions 2.7. For $E \in \mathcal{S}$, define $\Sigma E = E \wedge S^1$ and $\Omega E = F(S^1, E)$. Abbreviate $S = Q_\infty S^0$ and define the homotopy groups of E by

$$\pi_r E = \pi(\Sigma^r S, E) \text{ and } \pi_{-r} E = \pi(\Omega^r S, E) \quad \text{for } r \geq 0.$$

Since $\Sigma^r S \cong Q_\infty S^r$, by Lemma 2.2, $\pi_r E = \pi_r E_0$ for $r \geq 0$. A map $\theta: E \to E'$ in $\mathcal{S}$ is said to be a weak equivalence if $\pi_r \theta$ is an isomorphism for all integers r.

The adjunction between Σ and Ω gives natural maps

$$\eta: E \to \Omega\Sigma E \quad \text{and} \quad \varepsilon: \Sigma\Omega E \to E.$$

The following result is a version of the Puppe desuspension theorem [55].

Theorem 2.8. For all spectra E, η and ε are weak equivalences.

There is a category $H\mathcal{S}$ and a functor $L: h\mathcal{S} \to H\mathcal{S}$ such that L is the identity function on objects, L carries weak equivalences to isomorphisms, and L is universal with respect to the latter property [48, II and XI]. We call $H\mathcal{S}$ the stable homotopy category. Its morphisms are composites of morphisms in $h\mathcal{S}$ and formal inverses of weak equivalences in $h\mathcal{S}$. In $H\mathcal{S}$, the functors Σ and Ω become inverse equivalences of categories, and we therefore write $\Omega^r = \Sigma^{-r}$ for all integers r. It will be shown in [48] that $H\mathcal{S}$ has all the good properties one could hope for and, despite its wholly different definition, is in fact equivalent to the stable categories of Boardman [18] and Adams [7].

Let $H\mathcal{J}$ denote the category obtained from $h\mathcal{J}$ by formally inverting the weak equivalences and let $[Y, Z]$ denote the set of morphisms in $H\mathcal{J}$ or $H\mathcal{S}$ between spaces or spectra Y and Z. Again, we have

$$[X, E_0] \cong [Q_\infty X, E], \quad X \in \mathcal{J} \quad \text{and } E \in \mathcal{S}.$$

Q_∞ should be regarded as the stabilization functor from spaces to spectra. Let $\mathcal{V}$ denote the category of spaces in $\mathcal{T}$ of the (based) homotopy type of CW-complexes. For $X \in \mathcal{V}$ and $Y \in \mathcal{T}$, $[X, Y] = \pi(X, Y)$; the categories $h\mathcal{V}$ and $H\mathcal{T}$ are equivalent [48, III]. Analogous statements are valid for $H\mathcal{L}$ [48, XI].

$H\mathcal{L}$ admits a coherently associative, commutative, and unital smash product with unit S [48, XI]. Define a (commutative) ring spectrum to be a spectrum E together with an associative (and commutative) product $\phi: E \wedge E \to E$ with a two-sided unit $e: S \to E$. The following lemma will play a vital role in our study of Bott periodicity and Brauer lifting in VIII §2.

<u>Lemma 2.9.</u> The product ϕ of a ring spectrum E induces a map (again denoted ϕ) from $E \wedge E_0$ to E.

<u>Proof.</u> For spaces X, $E \wedge X$ is coherently naturally isomorphic to $E \wedge Q_\infty X$ [48, XI]; indeed, such a relationship between small and large smash products is a standard property of any good stable category. Via Definition 1.6, the identity map of E_0 determines a map $\psi: Q_\infty E_0 \to E$ of spectra. The required map is the composite

$$E \wedge E_0 \cong E \wedge Q_\infty E_0 \xrightarrow{1 \wedge \psi} E \wedge E \xrightarrow{\phi} E.$$

A spectrum E is said to be n-connected if $\pi_r E = 0$ for $r \leq n$ and to be connective if it is (-1)-connected. Infinite loop space theory is concerned with connective spectra, and we require the following observations.

<u>Lemma 2.10.</u> If C and D are $(q-1)$-connected and $\theta: D \to E$ is a map in $H\mathcal{L}$ such that $\pi_i\theta$ is an isomorphism for all $i \geq q$, then $\theta_*: [C, D] \to [C, E]$ is an isomorphism.

<u>Proof.</u> Let F denote the cofibre of θ. Up to sign, cofiberings and fiberings coincide in $H\mathcal{L}$ [48, XI], hence $\pi_i F = 0$ for $i \geq q$. By re-

placing C by a CW-spectrum, applying induction over its skeleta, and using the $\varprojlim^1$ exact sequence, we find easily that $[C, \Omega^i F] = 0$ for $i \geq 0$. The conclusion follows by the Barratt-Puppe sequence.

Lemma 2.11. For a spectrum E, there exists one and, up to equivalence, only one connective spectrum D and map $\theta: D \to E$ in $H\mathcal{J}$ such that $\pi_i\theta$ is an isomorphism for $i \geq 0$. If E is a ring spectrum, then D admits a unique structure of ring spectrum such that θ is a map of ring spectra.

Proof. While the existence and uniqueness could be proven by stable techniques, we simply note that the map $\tilde{\omega}: \Omega^\infty TE_0 \to E$ in $\mathcal{J}$ constructed in VII.3.2 has the properties required of θ and that, given $\theta: D \to E$ and $\theta': D' \to E'$ as specified, the naturality of $\tilde{\omega}$ yields the following commutative diagram in $H\mathcal{J}$, in which all arrows with targets other than E are isomorphisms:

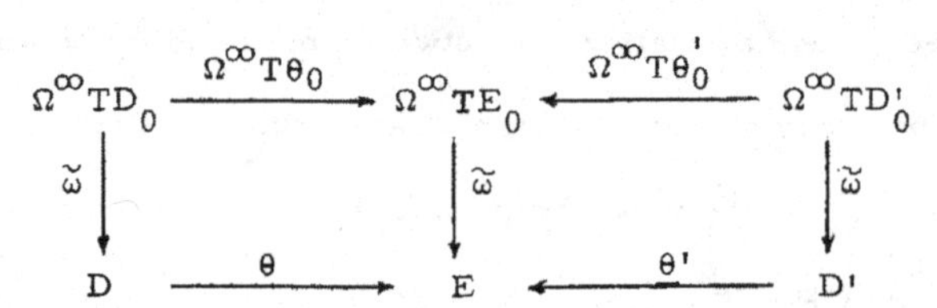

Since $D \wedge D$ is connective, the assertion about ring structures follows directly from the previous lemma.

Note that our proof not only gives an associated connective spectrum functor on $H\mathcal{J}$, it already gives such a functor on $\mathcal{J}$.

For a (commutative) ring R, the Eilenberg-MacLane spectrum $HR = \mathcal{K}(R, 0)$ is a (commutative) ring spectrum.

Lemma 2.12. If E is a connective ring spectrum, then the unique map $d: E \to H\pi_0 E$ in $H\mathcal{J}$ which realizes the identity map of $\pi_0 E$ is a map of ring spectra.

<u>Proof.</u> $\mathrm{Hom}(\pi_0 E, \pi_0 E) = H^0(E; \pi_0 E) = [E, H\pi_0 E]$ (see Definition 3.1), hence d is well-defined. d is a ring map because $\mathrm{Hom}(\pi_0 E \otimes \pi_0 E, \pi_0 E) = [E \wedge E, H\pi_0 E]$. Note that d can be explicitly constructed by application of the functor $\Omega^\infty T$ to the discretization $E_0 \to \pi_0 E_0 = \pi_0 E$.

Localizations and completions will often be needed in our work. Let T be a set of primes. Recall that an Abelian group A is said to be T-local if it is a module over the localization Z_T of Z at T and to be T-complete if $\mathrm{Hom}(Z[T^{-1}]/Z, A) = 0$ and the natural (connecting) homomorphism $A \to \mathrm{Ext}(Z[T^{-1}]/Z, A)$ is an isomorphism (where $Z[T^{-1}]$ is the localization of Z away from T). A (connected) simple space Y is said to be T-local or T-complete if each $\pi_i Y$ is T-local or T-complete. A localization $\lambda: X \to X_T$ or completion $\gamma: X \to \hat{X}_T$ of a simple space X at T is a map into a T-local or T-complete space such that

$$\lambda^*: [X_T, Y] \to [X, Y] \quad \text{or} \quad \gamma^*: [\hat{X}_T, Y] \to [X, Y]$$

is an isomorphism for all T-local or T-complete spaces Y or, equivalently, such that (with $Z_p = Z/pZ$)

$$\lambda_*: H_*(X; Z_T) \to H_*(X_T, Z_T) \quad \text{or each} \quad \gamma_*: H_*(X; Z_p) \to H_*(\hat{X}_T; Z_p), \quad p \in T,$$

is an isomorphism. λ and γ exist (and are unique), and $\hat{X}_T$ is equivalent to the completion of X_T at T and to $\times_{p \in T} \hat{X}_p$, where X_p and $\hat{X}_p$ denote the localization and completion of X at p. Localizations and completions commute with products, fibrations, and loops and localizations but not completions commute with wedges, cofibrations, suspensions, and smash products. However, the completion at T of $\gamma \wedge \gamma: X \wedge X' \to \hat{X}_T \wedge \hat{X}'_T$ is an equivalence.

The completions just described are those due to Bousfield and Kan [23]; the completions of Sullivan [73] are not adequate for our applications in VIII. A new topological treatment of localizations and completions will be given in [48], and it will be seen there that completions are not appreciably more difficult to analyze than localizations. Incidentally, the category $H\mathcal{J}$ (and not its equivalent $h\mathcal{V}$) is the appropriate one in which to study localizations and completions since Postnikov towers with infinitely many non-zero homotopy groups never have the homotopy type of CW-complexes.

We also need localizations and completions of connective spectra. This subject is nowhere treated in the literature, a situation that will be rectified in [48]. In $H\mathcal{J}$, the summary above applies verbatim with spaces replaced by connective spectra, the only twist being that the commutation of completions with products, fibrations, and loops implies their commutation with finite wedges, cofibrations, and suspensions, but still not with smash products. Again, the completions at T of $\gamma \wedge \gamma : E \wedge E' \to \hat{E}_T \wedge \hat{E}'_T$ and, for a simple space X, of $\gamma \wedge \gamma : E \wedge X \to \hat{E}_T \wedge \hat{X}_T$ are isomorphisms in $H\mathcal{J}$. In particular, the completion at T of a ring spectrum E is a ring spectrum with unit $S \xrightarrow{e} E \xrightarrow{\gamma} \hat{E}_T$ and product

$$\hat{E}_T \wedge \hat{E}_T \xrightarrow{\gamma} (\widehat{\hat{E}_T \wedge \hat{E}_T})_T \xrightarrow{(\widehat{\gamma \wedge \gamma})^{-1}} (\widehat{E \wedge E})_T \xrightarrow{\hat{\phi}} \hat{E}_T .$$

The zeroth space functor commutes with completions in the sense that the zeroth space of $\hat{E}_T$ is equivalent to the product of the completion at T of the component of the basepoint of E_0 and the discrete group $\mathrm{Ext}(Z[T^{-1}]/Z, \pi_0 E_0)$. When $\pi_0 E_0 = Z$, the latter group is the T-adic integers $\hat{Z}_T = \times_{p \in T} \hat{Z}_{(p)}$.

Finally, we shall need the following pair of results from [48]. Taken together, they assert that, under minimal technical hypotheses, a map between T-local infinite loop spaces which completes to an infinite loop map at each prime $p \in T$ is itself an infinite loop map.

Theorem 2.13. Let D and E be 0-connected spectra such that π_*D and π_*E are of finite type over Z_T and let $f: D_0 \to E_0$ be a map (in $H\mathcal{T}$) whose localization at p is the zeroth map of a map $\phi_p: D_p \to E_p$ (in $H\mathcal{A}$) for each prime $p \in T$. Assume either that D_0 and E_0 have no T-torsion and f is an H-map or that $D_0 \simeq \mathrm{Tel}(D_0^q)_T$ where each D_0^q is a finite CW-complex and $\varprojlim^1(\Sigma D_0^q, E_0) = 0$. Then f is the zeroth map of a map $\phi: D \to E$. Moreover, if $D \simeq \mathrm{Tel}\, D_T^q$ where each D^q is a finite CW-spectrum and $\varprojlim^1(\Sigma D^q, E) = 0$, then there is exactly one such map ϕ which localizes to the given map ϕ_p at each prime $p \in T$.

Theorem 2.14. Let D and E be 0-connected spectra such that π_*D and π_*E are of finite type over $Z_{(p)}$ and let $f: D_0 \to E_0$ be a map (in $H\mathcal{T}$) whose rationalization is an H-map and whose completion at p is the zeroth map of a map $\hat{\phi}_p: \hat{D}_p \to \hat{E}_p$ (in $H\mathcal{A}$). Assume either that D_0 and E_0 have no p-torsion and f is an H-map or that $D_0 \simeq \mathrm{Tel}(D_0^q)_p$ where each D_0^q is a finite CW-complex and $\varprojlim^1(\Sigma D_0^q, E_0) = 0$. Then f is the zeroth map of a map $\phi: D \to E$. Moreover, if $D \simeq \mathrm{Tel}\, D_p^q$ where each D^q is a finite CW-spectrum and $\varprojlim^1(\Sigma D^q, E) = 0$, then there is exactly one such map ϕ which completes at p to the given map $\hat{\phi}_p$.

One pleasant feature of these results is the complete irrelevance of $\varprojlim^1$ terms associated to the spaces D_n and E_n for $n > 0$. As will be discussed in [48], results of Anderson [9 and 11] show that the stable $\varprojlim^1$ terms vanish in the cases relevant to this book.

3. Cohomology; periodic spaces and spectra

Definition 3.1. For spectra Y and E, define the E-cohomology of Y by

$$E^nY = [\Omega^n Y, E] = [Y, \Sigma^n E].$$

For a space X, define $E^nX = E^nQ_\infty X$; this is what is usually called the reduced E-cohomology of X. In terms of $H\mathcal{J}$, E^nX can be described by

$$E^nX = [X, EV] \text{ if } \dim tV = n \geq 0 \quad \text{and} \quad E^{-n}X = [X, \Omega^n E_0] \text{ if } n \geq 0.$$

This description of E^nX will be essential to our treatment of orientation theory in the next chapter. A complete analysis of homology and cohomology theories within the framework given by $H\mathcal{J}$ will be presented in [48]. Suffice it to say that all of the familiar machinery is available.

We shall shortly need the following result, which is proven in [48]. Observe that, with the standard coordinatization $Y_i = YR^i$ and $tV = V \cup \infty$, the restriction of $\theta \in [Y, E]$ to maps $\theta_i \in [Y_i, E_i]$ specifies a homomorphism $E^0Y \to \varprojlim E^iY_i$.

Proposition 3.2. For all spectra Y and E, the map $E^0Y \to \varprojlim E^iY_i$ is an epimorphism with kernel isomorphic to $\varprojlim^1 E^{i-1}Y_i$.

This result is closely related to Whitehead's analysis of cohomology theories on spaces in terms of less stringent notions of spectra and their maps than we have been using. Define (coordinatized) weak prespectra by deleting the inclusion condition on the $\tilde{\sigma}_i$ in Definition 1.7, define weak Ω-prespectra by requiring the $\tilde{\sigma}_i$ to be weak equivalences, and define weak maps of weak prespectra by requiring only that $\theta_{i+1} \circ \sigma_i \simeq \sigma'_i \circ \Sigma\theta_i$ (and retain the term map for the case when equality holds). Weak Ω-prespectra and weak maps determine (additive) cohomology theories on spaces and morphisms thereof. Two weak maps determine the same morphism if $\theta_i = \theta'_i$ in $H\mathcal{J}$; we then say that θ and θ' are weakly homotopic. Similarly, we say that two maps $\theta, \theta' \in [Y, E]$ are weakly homotopic if $\theta V = \theta' V$ in $H\mathcal{J}$ for each indexing space V. In view of Theorem 1.8, we see that θ and θ' are weakly homotopic if and only if $\theta - \theta'$ is in the kernel $\varprojlim^1 E^{i-1}Y_i$ of the epimorphism $E^0Y \to \varprojlim E^iY_i$.

Let $WH\mathscr{S}$ denote the category of spectra and weak homotopy classes of maps in $H\mathscr{S}$.

Inductive mapping cylinder arguments [43, Theorem 4] allow one to replace a weak prespectrum T by a weakly equivalent (coordinatized) prespectrum, functorially up to weak homotopy. One can then use the functor Ω^{∞} of Definition 1.4 (on the coordinatized level) and the functor ψ of Theorem 1.8 to obtain an actual spectrum in $\mathscr{S}$. Alternatively, one can use Q_{∞} and a direct telescope construction on the spectrum level to pass from weak prespectra to spectra [48,XII]. Either way, one obtains the following result. For a spectrum E, we may of course regard the underlying coordinatized prespectrum $\nu\emptyset E$ as a weak Ω-prespectrum.

<u>Theorem 3.3.</u> There is a functor L from the category of weak prespectra and weak homotopy classes of maps to the category $WH\mathscr{S}$ and there is a map $\kappa: T \rightarrow \nu\emptyset LT$ of weak prespectra, natural up to weak homotopy, which is a weak equivalence if T is a weak Ω-prespectrum. Further, for $E \in \mathscr{S}$, there is a natural weak equivalence $\rho: L\nu\emptyset E \rightarrow E$ of spectra such that the following composite is the identity map:

$$\nu\emptyset E \xrightarrow{\kappa} \nu\emptyset L\nu\emptyset E \xrightarrow{\nu\emptyset\rho} \nu\emptyset E.$$

Finally, there is a natural weak equivalence $\omega: \Omega LT \rightarrow L\Omega T$ such that the following diagrams commute:

$$\begin{array}{ccc} \Omega T & \xrightarrow{\Omega\kappa} & \Omega\nu\emptyset LT \\ \kappa\downarrow & & \| \\ \nu\emptyset L\Omega T & \xleftarrow{\nu\emptyset\omega} & \nu\emptyset\Omega LT \end{array} \quad \text{and} \quad \begin{array}{ccc} L\nu\emptyset\Omega E & = & L\Omega\nu\emptyset E \\ \rho\downarrow & & \uparrow\omega \\ \Omega E & \xleftarrow{\Omega\rho} & \Omega L\nu\emptyset E \end{array}$$

Together with standard representability arguments, this result implies that a cohomology theory on spaces extends uniquely to a cohomology theory on spectra and that a morphism of cohomology theories on spaces extends to a morphism of cohomology theories on spectra but not, however, uniquely since there are non-trivial maps in $H\mathscr{S}$ which are weakly null homotopic and thus induce the trivial morphism of cohomology theories on spaces. Formally, $WH\mathscr{S}$ is equivalent to the category of cohomology theories on spaces and $H\mathscr{S} \to WH\mathscr{S}$ corresponds to the forgetful functor from cohomology theories on spectra to cohomology theories on spaces.

We also require the analogous result for products. Recall the notion of a pairing $(T', T'') \to T$ of weak prespectra from IX.2.5 below. (Our signs differ from Whitehead's [80] since we write suspension coordinates on the right.) It is easy to see that a map $\theta: E' \wedge E'' \to E$ of spectra determines a pairing $\pi(\theta): (\nu\emptyset E', \nu\emptyset E'') \to \nu\emptyset E$ of weak prespectra (compare IV.1.3 below). Via either of the two lines of proof of the previous theorem, one can verify the following addendum [48, XII].

Proposition 3.4. A pairing $f: (T', T'') \to T$ of weak prespectra determines a map $\emptyset(f): LT' \wedge LT'' \to LT$ of spectra, unique up to weak homotopy, such that $\kappa \circ f$ and $\pi(\emptyset(f)) \circ (\kappa, \kappa)$ are weakly homotopic pairings $(T', T'') \to LT$. If $\theta: E' \wedge E'' \to E$ is a map of spectra, then $\rho \circ \emptyset(\pi(\theta))$ is weakly homotopic to $\theta \circ (\rho \wedge \rho)$.

The notion of pairing gives rise to a notion of weak ring prespectrum, and this notion is adequate for the study of products in cohomology theories on spaces [80]. Define a weak ring spectrum in $H\mathscr{S}$ by only requiring the associativity and unit laws to hold up to weak homotopy. The proposition and theorem imply that a weak ring prespectrum T determines a weak ring spectrum LT.

The distinction between maps in $H\mathscr{S}$ and morphisms of cohomology theories on spaces and the concomitant distinction between weak ring spectra and ring spectra are folklore. The E_∞ ring spectra to be introduced in chapter IV are always honest ring spectra, and we shall construct E_∞ ring spectra from E_∞ ring spaces in chapter VII. Thus, where it applies, our work will circumvent any need for analysis of $\varprojlim^1$ terms. For the periodic K-theories, the relevant $\varprojlim^1$ terms vanish because $KU^{-1}(BG) = 0$ and $KO^{-1}(BG)$ is a finite dimensional vector space over Z_2 for any compact Lie group G [14]. For the connective K-theories, the relevant $\varprojlim^1$ terms vanish by results of Anderson [11]. We shall keep track of these distinctions in this section but, because of the arguments just given, shall generally ignore them in the rest of the book.

We now turn to the study of periodic spaces and spectra, and we fix an even positive integer d throughout the discussion. As a harmless simplification, we assume henceforward that the zeroth spaces of all spectra lie in the category $\mathcal{V}$ of spaces of the based homotopy type of CW-complexes.

Definition 3.5. For $\mathcal{C} = h\mathcal{V}$ or $\mathcal{C} = WH\mathcal{A}$, define $\Pi\mathcal{C}$, the category of periodic objects in $\mathcal{C}$, to be the category of pairs (X, χ) where $X \in \mathcal{C}$ and $\chi: X \to \Omega^d X$ is an isomorphism in $\mathcal{C}$. The morphisms $\zeta: (X, \chi) \to (X', \chi')$ are the maps $\zeta: X \to X'$ such that $\Omega^d \zeta \circ \chi = \chi' \circ \zeta$.

Proposition 3.6. The zeroth space functor from periodic spectra to periodic spaces is an equivalence of categories.

Proof. We shall work with coordinatized prespectra (as in Definition 1.7) taken with each B_i of dimension d and with tV the one-point compactification of V, so that σ_i maps $\Sigma^d T_i$ to T_{i+1} for all $i \geq 0$. Let $(X, \chi) \in \Pi h\mathcal{V}$. Let $X_i = X$, let $\chi_i = \chi: X_i \to \Omega^d X_i$, and let $\alpha_i: \Sigma^d X_i \to X_{i+1}$ be the adjoint of χ. Then $\{X_i, \alpha_i\}$ is a weak prespectrum and $\{\chi_i\}: \{X_i\} \to \Omega^d\{X_i\}$ is a weak map of weak prespectra (because, since d is even, the interchange of coordinates self homeomorphism of $\Omega^d\Omega^d X$ is homotopic to the identity). Define $KX = L\{X_i\} \in \mathcal{A}$ and define $\bar{\chi}: KX \to \Omega^d KX$ to be the composite of $L\{\chi_i\}$ and the natural isomorphism $\omega^{-1}: L\Omega^d\{X_i\} \to \Omega^d L\{X_i\}$ of Theorem 3.3. With the evident maps, we thus obtain a functor $K: \Pi h\mathcal{V} \to \Pi WH\mathcal{S}$. By the naturality of κ and the first diagram of Theorem 3.3, the zeroth map of $\kappa: \{X_i\} \to \nu\emptyset L\{X_i\}$ specifies an equivalence of periodic spaces $(X,\chi) \to (KX,\overline{\chi})$. Conversely, given $(E,\xi) \in \Pi WH\mathcal{S}$, write $(X,\chi) = (E_0, \xi_0)$, define $\gamma_0 = 1: E_0 \to X$ and define $\gamma_i: E_i = ER^{di} \to X$ inductively as the composite

$$E_i \xrightarrow{\xi_i} \Omega^d E_i \xrightarrow{\tilde{\gamma}_{i-1}^{-1}} E_{i-1} \xrightarrow{\gamma_{i-1}} X .$$

Then $\{\gamma_i\}: \nu\emptyset E \to \{X_i\}$ is a weak map and $\Omega^d\{\gamma_i\} \circ \{\xi_i\} = \{\chi_i\} \circ \{\gamma_i\}$ as weak maps. Define $\gamma: E \to KX$ to be the composite of $\rho^{-1}: E \to L\nu\emptyset E$ and $L\{\gamma_i\}$. By the naturality of ρ and the second diagram of Theorem 3.3, $\gamma: (E,\xi) \to (KX,\overline{\chi})$ is a weak equivalence of periodic spectra (and of course, as we have used several times, weak equivalences are isomorphisms in $WH\mathcal{S}$).

We are really interested not in periodic spectra but in "periodic connective spectra", and we write $\mathcal{S}_c$ for the category of connective spectra.

<u>Definition 3.7.</u> Define $\Pi WH\mathcal{S}_c$, the category of periodic connective spectra, to be the category of pairs (D,δ), where D is a connective spectrum and $\delta: D \to \Omega^d D$ is a map in $WH\mathcal{S}$ such that $\delta_0: D_0 \to \Omega^d D_0$ is an equivalence of spaces. The morphisms $\zeta: (D,\delta) \to (D',\delta')$ are the maps $\zeta: D \to D'$ such that $\Omega^d\zeta \circ \delta = \delta' \circ \zeta$ in $WH\mathcal{S}$.

<u>Proposition 3.8.</u> The associated connective spectrum functor from periodic spectra to periodic connective spectra is an equivalence of categories.

<u>Proof.</u> Given a periodic spectrum (E,ξ), let $\theta: D \to E$ be its associated connective spectrum of Lemma 2.11 and note that Lemma 2.10 gives a unique map $\delta: D \to \Omega^d D$ such that $\Omega^d\theta \circ \delta = \xi \circ \theta$ (since these results

for $H\mathcal{J}$ clearly remain valid for $WH\mathcal{J}$). Clearly $\theta_0: (D_0, \delta_0) \to (E_0, \xi_0)$ is an equivalence of periodic spaces. Given a periodic connective spectrum (D, δ), the last part of the previous proof applies verbatim to yield a map $\gamma: D \to KD_0$ such that $\Omega^d \gamma \circ \delta = \bar{\delta}_0 \circ \gamma$ in $WH\mathcal{J}$. By inspection of the zeroth space level, $\pi_i \gamma$ is an isomorphism for $i \geq 0$. Therefore γ induces a natural isomorphism between the identity functor of $\Pi WH\mathcal{J}_c$ and the composite

$$\Pi WH\mathcal{J}_c \to \Pi h\mathcal{V} \xrightarrow{K} \Pi WH\mathcal{J} \to \Pi WH\mathcal{J}_c .$$

The conclusion follows formally from Proposition 3.6.

The following consequence has already been used in chapter I and will be used more deeply in chapter VIII.

Corollary 3.9. Let (D, δ) and (D', δ') be periodic connective spectra and let $\lambda: (D_0, \delta_0) \to (D_0', \delta_0')$ be a map of periodic spaces. Then there is a unique map $\Lambda: (D, \delta) \to (D', \delta')$ of periodic connective spectra with zeroth map λ . If λ is an equivalence, then Λ is a weak equivalence of spectra.

Proof. Λ is given by Lemma 2.10 as the unique map (up to weak homotopy) such that the following diagram commutes in $WH\mathcal{J}$:

$$\begin{array}{ccc} D & \xrightarrow{\Lambda} & D' \\ \gamma\downarrow & & \downarrow\gamma \\ KD_0 & \xrightarrow{K\lambda} & KD_0' \end{array}$$

Note that, when λ is an equivalence, only one of δ_0 and δ_0' need be assumed to be an equivalence. The corollary characterizes the periodic connective spectrum associated to a periodic space. We also need a multiplicative elaboration applicable to periodic ring spaces.

Definition 3.10. A ring space is a space X together with a basepoint 0 and unit point 1, products $\oplus$ and $\otimes$, and an additive inverse map such that the ring axioms hold up to homotopy and 0 is a strict zero for $\otimes$ (so that $\otimes$ factors through $X \wedge X$); X is said to be commutative if $\otimes$ is homotopy commutative. By [26, I.4.6], X is equivalent to $X_0 \times \pi_0 X$ as an H-space under $\oplus$, where X_0 denotes the 0-component. Define $\Pi Rh\mathcal{V}$, the category of periodic ring spaces, to be the subcategory of $\Pi h\mathcal{V}$ whose objects are pairs (X, χ) such that X is a commutative ring space and $\chi: X \to \Omega^d X$ is adjoint to the composite

$$X \wedge S^d \xrightarrow{1 \wedge b} X \wedge X \xrightarrow{\otimes} X$$

for some $[b] \in \pi_d X = \pi_d(X_0, 0)$; the morphisms $\zeta: (X, \chi) \to (X', \chi')$ are the maps $\zeta: X \to X'$ of ring spaces such that $\zeta_*[b] = [b']$. Note that χ is automatically an H-map with respect to $\oplus$ and is determined by its restriction $X_0 \to \Omega_0^d X_0$ to basepoint components and by $\pi_0\chi: \pi_0 X \to \pi_0 \Omega^d X = \pi_d X$; since $(\pi_0\chi)[a] = [a][b]$ for $[a] \in \pi_0 X$, $\pi_d X$ must be the free $\pi_0 X$-module generated by $[b]$.

Definition 3.11. For $\mathcal{C} = WH\mathcal{S}$ or $\mathcal{C} = WH\mathcal{S}_c$, define $\Pi R\mathcal{C}$, the category of periodic ring objects in $\mathcal{C}$, to be the subcategory of $\Pi\mathcal{C}$ whose objects are the pairs (E, ξ) such that E is a weak commutative ring spectrum and $\xi: E \to \Omega^d E$ is adjoint to the composite

$$E \wedge S^d \xrightarrow{1 \wedge b} E \wedge E_0 \xrightarrow{\phi} E$$

for some $[b] \in \pi_d E_0$, where ϕ is as constructed in Lemma 2.9; the morphisms $\zeta: (E, \xi) \to (E', \xi')$ are the maps $\zeta: E \to E'$ of weak ring spectra such that $\zeta_*[b] = [b']$. By Lemmas 2.4 and 2.5, the zeroth space functor $\mathcal{C} \to h\mathcal{V}$ induces a functor $\Pi R\mathcal{C} \to \Pi Rh\mathcal{V}$.

We have the following complement to Proposition 3.6.

Proposition 3.12. The zeroth space functor from periodic ring spectra to periodic ring spaces is an equivalence of categories.

Proof. Given $(X, \chi) \in \Pi\mathrm{Rh}\mathcal{V}$, the following diagram is homotopy commutative (where b determines χ and τ denotes the transposition):

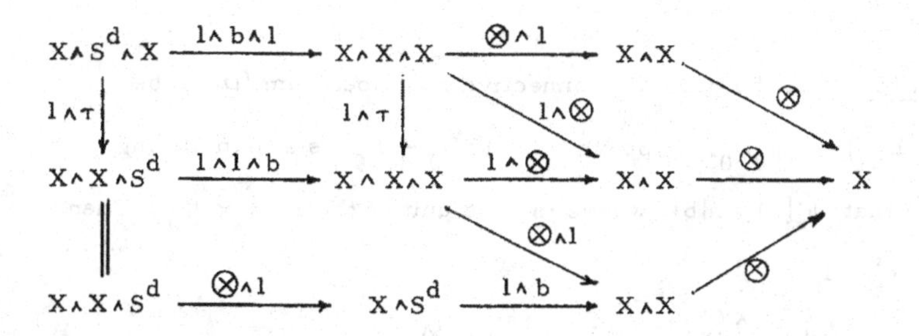

Therefore $\otimes: X_i \wedge X_j = X \wedge X \to X = X_{i+j}$ defines a pairing $(\{X_i\}, \{X_i\}) \to \{X_i\}$ and thus determines a weak ring spectrum structure on KX by Proposition 3.4. That result also implies that $\kappa_0: X \to K_0 X$ is a map of ring spaces. Conversely, given $(E, \xi) \in \Pi\mathrm{RWH}\mathcal{S}$, it is straightforward to verify that the map $\{\gamma_i\}: \nu\emptyset E \to \{X_i\}$, $X_i = E_0$, in the proof of Proposition 3.6 carries the pairing induced by the product on E to that just constructed. Therefore $\gamma: E \to KE_0$ is a map of weak ring spectra.

The analogous complement to Proposition 3.8 is a direct consequence of Lemma 2.11 (which clearly remains valid for weak ring spectra) and application of the last part of the argument just given to $(D, \delta) \in \Pi\mathrm{RWH}\mathcal{S}_c$.

Proposition 3.13. The associated connective spectrum functor from periodic ring spectra to periodic connective ring spectra is an equivalence of categories.

Corollary 3.14. Let (D, δ) and (D', δ') be periodic connective ring spectra and let $\lambda: (D_0, \delta_0) \to (D_0', \delta_0')$ be a map of periodic ring spaces. Then there is a unique map $\Lambda: (D, \delta) \to (D', \delta')$ of periodic connective ring spectra with zeroth map λ.

Proof. The maps $K\lambda$ and γ in the proof of Corollary 3.9 are maps of weak ring spectra, and it follows from Lemma 2.10 that Λ is so as well.

Remarks 3.15. Let the periodic connective ring spectrum (D, δ) be determined by $[b] \in \pi_d D_0$. Suppose that $\psi: D_0 \to D_0$ is a map of ring spaces such that $\psi_*[b] = n[b]$, where n is a unit of the ring $\pi_0 D_0$. Then the adjoint δ_0' of

$$D_0 \wedge S^d \xrightarrow{1 \wedge (\frac{1}{n})b} D_0 \wedge D_0 \xrightarrow{\otimes} D_0$$

is an equivalence. Since $\psi_*(\frac{1}{n}[b]) = [b]$, the corollary yields a map $\Psi: (D, \delta') \to (D, \delta)$ of periodic connective ring spectra with zeroth map ψ.

III. Orientation theory

The notion of orientability with respect to an extraordinary cohomology theory is central to bundle theory. We shall here use the coordinate-free spectra of chapter II to relate orientation theory to the geometric classifying spaces of chapter I. We shall think of a monoid-valued $\mathcal{J}_*$-functor G which maps to F as specifying a theory of sphere bundles (orthogonal, topological, etc.). In section 1, we shall use the general theory of fibrations developed in [47] to make rigorous a folklore treatment of orientation theory for GV-bundles oriented with respect to a commutative ring spectrum E. In section 2, we shall use the two-sided bar construction discussed in I §2 to give a precise geometric description of a classifying space $B(GV;E)$ for E-oriented GV-bundles. That it does indeed classify will be deduced from the much more general classification theorems for fibrations and bundles with additional structure established in [47], and several other consequences of the general theorems there will also be discussed.

1. Elementary orientation theory.

It is folklore that the theory of Thom complexes and orientations works particularly smoothly if one starts with spherical fibrations $\xi : D \to X$ with a given cross-section $\sigma : X \to D$ such that σ is a cofibration. One then defines the Thom complex $T\xi$ to be the quotient space D/X. For an n-plane bundle, the idea is to think of the n-sphere bundle obtained by applying one-point compactification to each fibre. The Thom complex obtained in this way will usually agree with that obtained by one-point compactification of the total space and will always agree with that obtained by taking the quotient of the unit disc bundle by its boundary (n-1)-sphere bundle.

Clearly, if the homotopy type of $T\xi$ is to be an invariant of the fibre homotopy equivalence class of ξ, then the latter notion must be defined in terms of section preserving fibrewise maps (and homotopies). In turn, if homotopic maps $X' \to X$ are to induce fibre homotopy equivalent fibrations from ξ, then the covering homotopy property must also be formulated in terms of section preserving fibrewise maps. It is then not immediately obvious how much of the standard theory of fibrations goes through; for example, the usual procedure for replacing a spherical quasifibration by a spherical fibration clearly fails.

In [47, §1-3], the basic theory of fibrations is redeveloped with fibres and maps of fibres constrained to lie in any preassigned category of spaces $\mathcal{F}$. Let V be a finite dimensional real inner product space and take $\mathcal{F}$ to be the subcategory of $\mathcal{J}$ which consists of the spaces of the (based) homotopy type of tV and their (based) homotopy equivalences. The basepoints of fibres are required to define cross-sections which are fibrewise cofibrations (see [47, 5.2]). This condition both allows our proposed construction of Thom complexes and is necessary to circumvent the problem with quasifibrations mentioned above [47, § 5]. We shall call spherical fibrations of the sort just specified "FV-bundles".

Now assume given a grouplike monoid-valued $\mathcal{J}_*$-functor G together with a morphism $G \to F$ of monoid-valued $\mathcal{J}_*$-functors, as in I.2.5. Define a GV-bundle to be an FV-bundle together with a reduction of its structural monoid to GV. The precise meaning of a "reduction" in this generality is specified in [47, 10.4], and the cited definition shows that GV-bundles are naturally equivalent to Steenrod fibre bundles with group GV and fibre tV when G is group-valued and GV acts effectively on tV.

In this context, fibrewise joins are replaced by fibrewise smash products. Explicitly, if ξ and ψ are GV and GW bundles over X and Y with total spaces $D\xi$ and $D\psi$, define $\xi \wedge \psi$ to be the $G(V \oplus W)$-bundle over $X \times Y$ with total space $D(\xi \wedge \psi) = (D\xi \times D\psi)/(\approx)$, where the equivalence identifies the wedge $(\sigma x, \psi^{-1} y) \vee (\xi^{-1} x, \sigma y)$ to the point $(\sigma x, \sigma y)$ for each $(x, y) \in X \times Y$. The projection $\xi \wedge \psi$ is induced from $\xi \times \psi$ and the section is induced from $\sigma \times \sigma$. $\xi \wedge \psi$ is in fact a $G(V \oplus W)$-bundle because it is an $F(V \oplus W)$-bundle by [47, 5.6] and because it inherits a reduction from those of ξ and ψ by [47, 5.6 and 10.4] and use of the Whitney sums given in I. 2. 2 on the bar constructions which appear in [47, 10.4]. We have an evident homeomorphism

$$T\xi \wedge T\psi \xrightarrow{\cong} T(\xi \wedge \psi) .$$

When $X = Y$, define the Whitney sum $\xi \oplus \psi$ to be the $G(V \oplus W)$-bundle over X induced from $\xi \wedge \psi$ by the diagonal map $\Delta: X \to X \times X$. We then have a homeomorphism

$$T(\xi \oplus \psi) \xrightarrow{\cong} T\xi^*(\psi)/T\psi ,$$

where $\xi^*(\psi)$ is the GW-bundle over $D\xi$ induced from ψ by $\xi: D\xi \to X$; of course, ψ is the GW-bundle over X induced from $\xi^*(\psi)$ by $\sigma: X \to D\xi$ (since $\xi\sigma = 1$), and the GW-bundle map over σ induces the inclusion used to define the quotient on the right.

Let $E \in \mathcal{J}$ be a commutative ring spectrum and recall the definition of E^*X from II. 3. 1.

<u>Definition 1.1</u>. A GV-bundle ξ is E-orientable if there exists a class $\mu \in E^n T\xi$, $n = \dim tV$, such that μ restricts to a generator of the free $\pi_* E$-module $E^* T\chi$ for each fibre χ of ξ (where fibres are thought of as GV-bundles over points of the base space).

<u>Remarks 1.2.</u> Let $\theta: D \to E$ be a map of commutative ring spectra. Clearly ξ is E-orientable if it is D-orientable. Conversely, if D is connective and

$\pi_i\theta$ is an isomorphism for $i \geq 0$, then ξ is D-orientable if it is E-orientable (because $\theta_*: D^n T\xi \to E^n T\xi$ is an isomorphism since $T\xi$ is (n-1)-connected). By II. 2.11 and 2.12, it follows that orientation theory depends only on connective spectra and that a bundle ξ is $H\pi_0 E$-orientable if it is E-orientable.

Henceforward, write R for $\pi_0 E = \pi_0 E_0$. Recall that $(HR)^*(X)$ is the ordinary reduced cohomology $\tilde{H}^*(X,R)$. By an R-orientation (or orientation if R = Z) of a GV-bundle ξ, we understand a class $\mu \in \tilde{H}^n(T\xi; R)$ such that μ restricts to a generator of the free R-module $\tilde{H}^n(T\chi; R)$ for each fibre χ; the pair (ξ, μ) is then said to be an R-oriented GV-bundle. Since we can identify $\tilde{H}^n(T\chi; \pi_* E)$ with $E^* T\chi$, μ restricts to a definite fundamental class in $E^n T\chi$ for each fibre χ.

Definition 1.3. An E-orientation of an R-oriented GV-bundle ξ is a class $\mu \in E^n T\xi$, $n = \dim tV$, such that μ restricts to the fundamental class of $E^n T\chi$ for each fibre χ; the pair (ξ, μ) is then said to be an E-oriented GV-bundle.

Thus E-orientations are required to be consistent with preassigned R-orientations. The following proof of the Thom isomorphism theorem should help motivate this precise definition. Let X^+ denote the union of X and a disjoint basepoint.

Theorem 1.4. Let (ξ, μ) be an E-oriented GV-bundle over a finite dimensional CW-complex X. Then the cup product with μ defines an isomorphism $E^* X^+ \to E^* T\xi$. Therefore $E^* T\xi$ is the free $E^* X^+$-module generated by μ.

Proof. The cup product is determined by the reduced diagonal $T\xi \to X^+ \wedge T\xi$ (which is induced via $\xi \wedge 1$ from the ordinary diagonal $D\xi^+ \to D\xi^+ \wedge D\xi^+$ of the total space). Now $\cup\mu$ induces a morphism of Atiyah-Hirzebruch spectral sequences which, on the E_2-level, is the isomorphism

$$\cup\mu : \tilde{H}^*(X^+; \pi_* E) \to \tilde{H}^*(T\xi; \pi_* E)$$

determined by the preassigned R-orientation of ξ.

Of course, the finite-dimensionality of X serves only to ensure convergence of the spectral sequences.

The following remarks summarize other basic facts about orientations; the proofs are immediate from the definitions, the previous theorem, and the facts about Thom complexes recorded above.

Remarks 1.5. Let X and Y be (finite-dimensional) CW-complexes.

(i) The trivial GV-bundle $\varepsilon = \varepsilon V : X \times tV \to X$ satisfies $T\varepsilon = X^+ \wedge tV$. The image under suspension of $1 \in E^0 X^+$ is an E-orientation of ε; it is called the canonical orientation and is denoted μ_0.

(ii) If (ψ, ν) is an E-oriented GV-bundle over Y and $f: X \to Y$ is a map, then $(Tf)^*(\nu)$ is an E-orientation of $f^*(\psi)$, where $Tf: Tf^*\psi \to T\psi$ is the induced map of Thom complexes. If, further, f is a cofibration, then the cup product with ν induces an isomorphism

$$E^*(Y/X) \xrightarrow{\cong} E^*(T\psi / T(\psi | X))$$

(by the long exact cohomology sequences and the five lemma).

(iii) If (ξ, μ) and (ψ, ν) are E-oriented GV and GW-bundles over X and Y, then $(\xi \wedge \psi, \mu \wedge \nu)$ is an E-oriented $G(V \oplus W)$-bundle over $X \times Y$, where $\mu \wedge \nu$ is the image of $\mu \otimes \nu$ under the external product [48, XII]

$$E^* T\xi \otimes E^* T\psi \longrightarrow E^*(T\xi \wedge T\psi) = E^* T(\xi \wedge \psi).$$

When X = Y, $\mu \oplus \nu$ denotes the induced E-orientation $(T\Delta)^*(\mu \wedge \nu)$ of $\xi \oplus \psi$.

(iv) If (ψ, ν) and $(\xi \oplus \psi, \omega)$ are E-oriented GW and $G(V \oplus W)$-bundles over X, where ξ is a GV-bundle over X, then the image μ of 1 under the composite isomorphism

$$E^*X^+ \xrightarrow{\cup\omega} E^*T(\xi \oplus \psi) = E^*(T\xi^*(\psi)/T\psi) \xrightarrow{(\cup\nu)^{-1}} E^*(D\xi/X) = E^*T\xi .$$

is the unique E-orientation of ξ such that $\mu \oplus \nu = \omega$.

(v) If ξ and ψ are stably equivalent GV and GW-bundles over X, so that $\xi \oplus \varepsilon(W \oplus Z)$ is equivalent to $\psi \oplus \varepsilon(V \oplus Z)$ for some Z, then ξ is E-orientable if and only if ψ is E-orientable.

2. Classification of E-oriented GV-bundles

We retain the notations of the previous section and assume that all spaces in sight are in the category $\mathscr{W}$ of spaces of the homotopy type of CW-complexes. By [46,A.6], B(Y,G,X) is in $\mathscr{W}$ if Y, G, and X are in $\mathscr{W}$.

Let SGV denote the component of the identity element of GV.

Let FR denote the group of units of the ring $R = \pi_0 E_0$ and let $FE \subset E_0$ denote the union of the corresponding components. Define $d: FE \rightarrow \pi_0 FE = FR$ to be the discretization map. Let $SFE \subset FE$ denote the component corresponding to the identity element of R. When E is the sphere spectrum $Q_\infty S^0$, the space $E_0 = QS^0$ coincides with $\tilde{F}$; in particular, FE = F and SFE = SF. In the general case, we may take the unit $e: Q_\infty S^0 \rightarrow E$ to be an honest map in $\mathscr{S}$ rather than just a map in $H\mathscr{S}$ (by II.2.7), and we also write e for the composite

$GV \rightarrow FV \subset F \xrightarrow{e} FE, \quad V \subset R^\infty.$

By the definition, II.1.1, of a spectrum, we have a homeomorphism $\tilde{\sigma}: E_0 \rightarrow F(tV, EV)$ for each finite-dimensional sub inner product space V of R^∞. We restrict attention to such V, and we identify FE with a subspace of F(tV,EV) via $\tilde{\sigma}$. We are given a morphism of monoids $GV \rightarrow FV \subset F(tV, tV)$, and composition of maps defines a right action of GV on FE and of SGV on SFE. Define

$$B(GV; E) = B(FE, GV, *) \text{ and } B(SGV; E) = B(SFE, SGV, *).$$

We then have the following commutative "orientation diagram", in which the maps i are the evident inclusions and $B(GV; R)$ is defined to be $B(FR, GV, *)$:

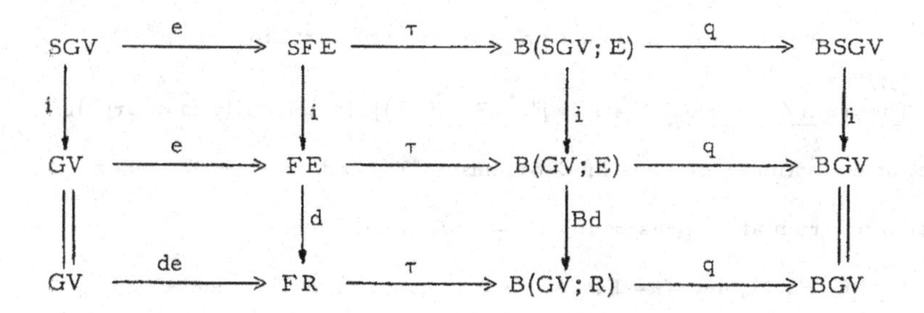

The rows are quasifibration sequences by [47, 7.9], and the maps q can be thought of as universal determinant bundles.

We shall interpret geometrically the functors and natural transformations represented on the homotopy category $h\mathcal{W}$ by the spaces and maps of the orientation diagram by quoting appropriate results of [47]. Recall that brackets denoted unbased homotopy classes in that paper but denote based homotopy classes here.

We note first that, by [47, 9.8], $[X^+, BFV]$ is naturally isomorphic to the set of equivalence classes of FV-bundles over X. Next, by [47, 11.1 and 10.4], $[X^+, B(GV\backslash FV, FV, *)]$ is naturally isomorphic to the set of equivalence classes of FV-bundles over X with a reduction of the structural monoid to GV. Here, by [47, 8.9], $B(GV\backslash FV, FV, *)$ is homotopy equivalent to BGV and the maps

$$q: B(GV\backslash FV, FV, *) \to BFV \quad \text{and} \quad Bj: BGV \to BFV$$

can be used interchangeably, $j: G \to F$. Of course, if G is group-valued and GV acts effectively on tV, then BGV also classifies Steenrod fibre bundles over X with group GV and fibre tV [47, 9.10]; here the map Bj induces the transformation obtained by sending a fibre bundle to its underlying spherical fibration, whereas q induces the transformation from GV-bundles to FV-bundles obtained by forgetting the reduction of the structural monoid. Given $i: H \to G$, BHV can be regarded as classifying either FV-bundles or GV-bundles with a reduction of their structural monoid to HV.

Theorem 2.1. For $X \in \mathcal{W}$, $[X^+, B(GV; E)]$ is naturally isomorphic to the set of equivalence classes of E-oriented GV-bundles over X under the relation of orientation preserving GV-bundle equivalence.

Proof. First, let G = F. An orientation μ of an FV-bundle $\xi: D \to X$ can be described as a homotopy class of maps $D \to EV$ such that, for any map $\psi: tV \to D$ which is a based homotopy equivalence into some fibre, the composite $\mu\psi: tV \to EV$ lies in FE. Here the basepoints of fibres determine the cross-section of ξ, and μ factors through $T\xi$ because $\mu\chi$ is a based map for each inclusion $\chi: \xi^{-1}(x) \subset D$. The condition $\mu\chi \in FE$ also ensures that the restriction of $\mu \in E^n T\xi$ to $E^n T\chi$ is a generator of $E^* T\chi$ over $\pi_* E$. In the language of [47; 10.1, 10.2, and 10.6], μ is an FE-structure defined with respect to the admissible pair (FE, EV). Therefore the result for G = F is a special case of [47, 11.1]. For general G, an orientation of a GV-bundle depends only on the underlying FV-bundle (and not on the reduction), hence the result for F implies the result for G by [47, 11.3]. Alternatively, when G is group-valued and GV acts effectively on tV, we could appeal to the bundle-theoretic result [47, 11.4] rather than to the quoted fibration-theoretic results.

The proofs of [47, 11.1 and 11.4] give explicit universal E-oriented GV-bundles (π, θ) with base B(GV; E) and with π classified by

$q: B(GV; E) \to BGV$ [47, 11.2]. Therefore q induces the obvious forgetful transformation from E-oriented GV-bundles to GV-bundles.

If FE happens to admit a structure of topological monoid such that $e: GV \to FE$ is a map of monoids, so that $B(GV; E) = FE/GV$ is homotopy equivalent to the fibre of $Be: BGV \to BFE$, then $\alpha^*(Be) \in [X^+, BFE]$ is the only obstruction to the E-orientability of the GV-bundle classified by $\alpha: X \to BGV$. As far as I know, the only examples are $E = HR$ and $E = S$. (both discussed below). Nevertheless, a similar obstruction will be constructed much more generally in the next chapter.

<u>Example 2.2.</u> Let $E = HR$. It is not hard to construct a model for HR such that $e: GV \to FHR$ is a morphism of monoids. Rather than give the details, we note that $Bd: B(GV; HR) \to B(GV; R)$ is a homotopy equivalence, since $d: FHR \to FR$ is, and we can thus use the middle and bottom rows of the orientation diagram interchangeably. Clearly $de: GV \to FR$ is a morphism of monoids, $B(GV; R)$ is equivalent to the fibre of $B(de)$, and $B(de)$ factors through $Bd: BGV \to B\pi_0 GV$. $BSGV$ is contained in $B(GV; R)$, and we have the following commutative diagram:

$$\begin{array}{ccccc} BSGV & \xrightarrow{i} & BGV & \xrightarrow{Bd} & K(\pi_0 GV, 1) \\ \downarrow & & \| & & \downarrow \\ B(GV; R) & \xrightarrow{q} & BGV & \xrightarrow{B(de)} & K(FR, 1) \end{array}$$

The GV-bundle classified by $\alpha: X \to BGV$ is R-orientable if and only if $\alpha^* B(de) = 0$, and this holds if $\pi_0 GV = \{1\}$ or if $\pi_0 GV = Z_2$ and either char $R = 2$ or the first Stiefel-Whitney class $w_1 = \alpha^* Bd \in H^1(X; Z_2)$ is zero. By the diagram, $BSGV \to B(GV; R)$ is a homotopy equivalence if $\pi_0 GV = FR$.

Let (π_0, θ_0) be the (integrally) oriented GV-bundle classified by the inclusion of BSGV in B(GV; Z). Via $\alpha \to \alpha^*(\pi_0, \theta_0)$, $[X^+, BSGV]$ is naturally isomorphic to the set of equivalence classes of SGV-bundles, by which we understand GV-bundles with a canonical orientation. In other words, Theorem 2.1 allows us to choose compatible preferred orientations simultaneously on all GV-bundles with a reduction of their structural monoid to SGV.

<u>Definition 2.3.</u> An E-oriented SGV-bundle is an E-oriented GV-bundle and an SGV-bundle such that the preassigned R-orientation is that induced from the canonical orientation.

The map $Bd: B(GV; E) \to B(GV; R)$ induces the evident forgetful transformation from E-oriented to R-oriented GV-bundles and is an isomorphism on components. Our original definition of an E-orientation takes cognizance of the fact that the set of equivalence classes of E-oriented GV-bundles over X is the union of the inverse images under $(Bd)_*$ of the elements of the set of R-oriented GV-bundles over X. The image of B(SGV; E) under $Bd \circ i$ is precisely $BSGV \subset B(GV; R)$. This implies the following corollary.

<u>Corollary 2.4.</u> For $X \in \mathscr{W}$, $[X^+, B(SGV; E)]$ is naturally isomorphic to the set of equivalence classes of E-oriented SGV-bundles over X.

We complete the analysis of the upper two rows of the orientation diagram in the following remarks (compare [47, 11.2]).

<u>Remarks 2.5.</u> (i) $[X^+, FE]$ is isomorphic to the set of E-orientations of the trivial GV-bundle εV over X. Indeed, given $\alpha: X \to FE$, its adjoint $X \times tV \to EV$ gives the corresponding E-orientation μ_α. α has image in SFE if and only if μ_α restricts to the canonical fundamental class of each fibre of εV.

(ii) The maps τ of the orientation diagram induce the transformations which send an E-orientation μ of εV to the equivalence class of the pair $(\varepsilon V, \mu)$.

(iii) $[X^+, GV]$ is isomorphic to the set of homotopy classes of GV-bundle equivalences $\varepsilon V \to \varepsilon V$. Indeed, given $\alpha: X \to GV$, its adjoint $X \times tV \to tV$ gives the second coordinate of the corresponding GV-bundle map ν_α. α has image in SGV if and only if $T\nu_\alpha$ preserves the canonical orientation of εV.

(iv) The maps e of the orientation diagram induce the transformations which send a GV-bundle map ν to the E-orientation $(T\nu)^*(\mu_0)$, where μ_0 is the canonical E-orientation of εV.

There is an analogy between orientations and trivializations that plays an important role in the applications. Suppose given morphisms $H \to G \to F$ of monoid-valued $\mathscr{I}_*$-functors, where H is group-valued. By [47, 10.3 and 11.1] (and, for cases such as H = Spin where HV fails to act effectively on tV, [47, 10.4 and 11.3]), $[X^+, GV/HV]$ is naturally isomorphic to the set of equivalence classes of GV-trivialized HV-bundles over X.

Remarks 2.6. $Be: GV/HV = B(GV, HV, *) \to B(FE, HV, *) = B(HV; E)$ induces the transformation which sends (ξ, ζ) to $(\xi, (T\zeta)^*(\mu_0))$, where μ_0 is the canonical E-orientation of the trivial GV-bundle. The following diagram commutes:

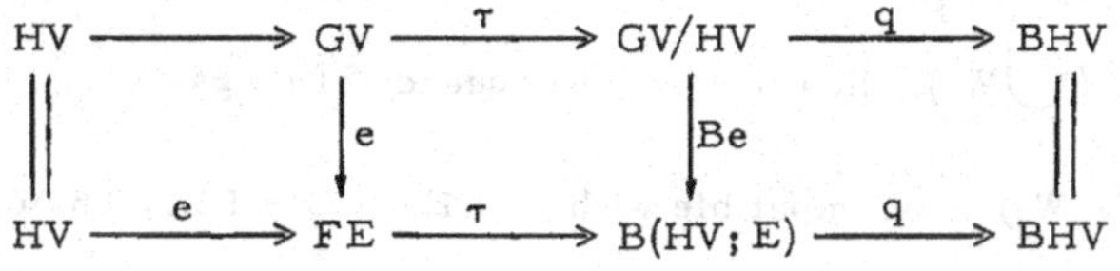

$$\begin{array}{ccccccc}
HV & \longrightarrow & GV & \xrightarrow{\tau} & GV/HV & \xrightarrow{q} & BHV \\
\| & & \downarrow{\scriptstyle e} & & \downarrow{\scriptstyle Be} & & \| \\
HV & \xrightarrow{e} & FE & \xrightarrow{\tau} & B(HV; E) & \xrightarrow{q} & BHV
\end{array}$$

In the upper row, τ induces the transformation which sends $\nu: \varepsilon V \to \varepsilon V$ to $(\varepsilon V, \nu)$ and q induces the evident forgetful transformation.

Example 2.7. Let E = S, so that FE = F and $B(GV; E) = F/GV$. Since $Be: BGV \to BF$ induces the transformation which sends a GV-bundle ξ to its stable fibre homotopy equivalence class, this class is the obstruction to the S-orientability of ξ. This fact can also be seen directly since, if

$V = R^m$ and $tV = S^m$, an S-orientation $\mu: T\xi \to QS^m$ has adjoint a map $D\xi \times S^n \to S^{n+m}$ for n suitably large (if the base of ξ is compact), and the latter map is the second coordinate of a stable trivialization of ξ.

Finally, we relate Theorem 2.1 to fibrewise smash products and discuss its naturality in E.

Remarks 2.8. Let V and W be orthogonal finite dimensional sub inner product spaces of R^∞. It is an easy consequence of the definition of the smash product on $H\mathscr{S}$ given in [48] that the product on E determines maps $\phi: FE \times FE \to FE$ and $\phi: EV \wedge EW \to E(V+W)$ (depending on an appropriate linear isometry $R^\infty \oplus R^\infty \to R^\infty$) such that the following diagram is commutative:

$$\begin{array}{ccccc} FE \times FE & \xrightarrow{\tilde{\sigma} \times \tilde{\sigma}} & F(tV, EV) \times F(tW, EW) & \xrightarrow{\wedge} & F(tV \wedge tW, EV \wedge EW) \\ \downarrow \phi & & & & \downarrow F(1,\phi) \\ FE & & \xrightarrow{\tilde{\sigma}} & & F(t(V+W), E(V+W)) \end{array}$$

(at least if the product on E is given by an honest map in $\mathscr{S}$; possible required formal inverses of weak equivalences would mildly complicate the argument to follow). If $\{V_i\}$ and $\{W_i\}$ are expanding sequences such that $V_i \perp W_i$ and $R^\infty = (\bigcup_i V_i) + (\bigcup_i W_i)$, then there is a sequence of maps $\phi_i: EV_i \wedge EW_i \to E(V_i + W_i)$ so compatible with $\phi: FE \times FE \to FE$. The diagram and the definitions of ω on F and of the right actions γ by the GV on FE imply that the following diagram also commutes:

$$\begin{array}{ccccc} FE \times GV \times FE \times GW & \xrightarrow{1 \times t \times 1} & FE \times FE \times GV \times GW & \xrightarrow{\phi \times \omega} & FE \times G(V+W) \\ \downarrow \gamma \times \gamma & & & & \downarrow \gamma \\ FE \times FE & & \xrightarrow{\phi} & & FE \end{array}$$

Therefore, since the bar construction commutes with products, ϕ and ω

induce

$$B(\omega;\emptyset): B(GV;E)\times B(GW;E) \to B(G(V+W);E).$$

Let (ξ,μ) and (ψ,ν) be E-oriented GV and GW-bundles over X and Y classified by $\overline{\alpha}$ and $\overline{\beta}$. Then $(\xi\wedge\psi,\mu\wedge\nu)$ is classified by $B(\omega;\emptyset)(\overline{\alpha}\times\overline{\beta})$. The proof is based on the observation that [47; 5.6, 7.4, and 11.1] imply that $B(\omega;\emptyset)$ can be covered by a map of E-oriented $G(V+W)$-bundles from the fibrewise smash product of the universal bundles over $B(GV;E)$ and $B(GW;E)$ to the universal bundle over $B(G(V+W);E)$.

<u>Remarks 2.9.</u> Let $\zeta: E\to E'$ be a map of commutative ring spectra. By the definition of a map of spectra, II. 1.1, and of the actions γ, the following diagrams are commutative:

$$\begin{array}{ccc} FE & \xrightarrow{\tilde{\sigma}} & F(tV,EV) \\ \downarrow{\scriptstyle\zeta} & & \downarrow{\scriptstyle F(1,\zeta)} \\ FE' & \xrightarrow{\tilde{\sigma}} & F(tV,E'V) \end{array} \qquad \text{and} \qquad \begin{array}{ccc} FE\times GV & \xrightarrow{\zeta\times 1} & FE'\times GV \\ \downarrow{\scriptstyle\gamma} & & \downarrow{\scriptstyle\gamma} \\ FE & \xrightarrow{\zeta} & FE' \end{array}$$

Therefore ζ induces a map $B\zeta: B(GV;E)\to B(GV;E')$, and the following diagram is obviously commutative:

$$\begin{array}{ccccc} FE & \xrightarrow{\tau} & B(GV;E) & \xrightarrow{q} & BGV \\ \downarrow{\scriptstyle\zeta} & & \downarrow{\scriptstyle B\zeta} & & \| \\ FE' & \xrightarrow{\tau} & B(GV;E') & \xrightarrow{q} & BGV \end{array}$$

The construction of the universal E and E' oriented GV-bundles (π,θ) and (π',θ') in [47, 11.1] shows that the map of Thom complexes $T\zeta: T\pi\to T\pi'$ induced by $B\zeta$ is such that the diagram

$$\begin{array}{ccc} T\pi & \xrightarrow{T\zeta} & T\pi' \\ \downarrow{\scriptstyle\theta} & & \downarrow{\scriptstyle\theta'} \\ EV & \xrightarrow{\zeta} & E'V \end{array}$$

is homotopy commutative. We conclude that $B\zeta$ induces the transformation which sends an E-oriented GV-bundle (ξ, μ) to the E'-oriented GV-bundle $(\xi, \zeta\mu)$, where we have also written ζ for the cohomology operation it determines.

IV. E_∞ ring spectra*

In the previous chapter, the basics of orientation theory were developed for a cohomology theory represented by a commutative ring spectrum E. In order to analyze the obstruction to orientability, and for many other purposes, it is desirable to have a more structured notion of a ring spectrum.

To see what is wanted, consider the unit space $FE \subset E_0$ and the component SFE of the identity element of $\pi_0 E$. The product $\phi: E \wedge E \to E$ and unit $e: S \to E$ determine H-space structures on FE and SFE. Recall that, when $E = S$, $FE = F$ and $SFE = SF$. When $E = kO$, Adams pointed out in [4, §7] that the group of kO-oriented spherical fibrations over X ought to play a key role in the analysis of J(X) and that the obstruction to kO-orientability ought to be directly related to the d-invariant. Now the d-invariant can be thought of as induced from the H-map $e: SF \to BO_\otimes = SFkO$, and Sullivan pointed out in [72, §6] that if e were to admit a delooping $Be: BSF \to BBO_\otimes$, then the fibre of Be ought to be equivalent to $B(SF; kO)$ and Be therefore ought to be the universal obstruction to kO-orientability.

Thus one wants at least sufficient structure on E to ensure that FE admits a classifying space (or delooping). One's first thought is to insist that FE admit a structure of topological monoid. One cannot require ϕ to be associative and unital, without passage to homotopy, since the smash product of spectra is itself only associative and unital after passage to homotopy. However, one can ignore the smash product, revert to Whitehead's notion of a ring spectrum defined in terms of a pair-

---

* (by J.P. May, F. Quinn, and N. Ray)

ing of spectra [80], and assume that the given pairing is strictly associative and unital. This is perfectly satisfactory in theory, and has been used by Patterson and Stong in an investigation of the orientability of bundles [52], but is at present of little use in practice since the only known strictly associative ring spectra are S (under the composition pairing) and the Eilenberg-Mac Lane spectra HR, where R is a ring.

We shall define the notion of an E_∞ ring spectrum in section 1. When E is such a spectrum, FE and SFE will not only have deloopings, they will actually be infinite loop spaces. Paradoxically, although the implications of an E_∞ ring structure are thus much stronger than the implications of strict associativity, it is not hard to construct E_∞ ring spectra. Indeed, among other examples, we shall see in section 2 that the various Thom spectra MG, including such exotic ones as $MTop$ and MF, are E_∞ ring spectra as they occur in nature. In VII, the first author will give a machine for the construction of E_∞ ring spectra from spaces with appropriate internal structure. This machine will construct E_∞ ring spectra which represent the various connective K-theories (geometric and algebraic) and the ordinary cohomology theories.

In section 3, we study E-orientation theory when E is an E_∞ ring spectrum. Here the obstruction to E-orientability takes on a conceptual form: it is a degree one "Stiefel-Whitney" class in the cohomology theory determined by the infinite loop space FE. We shall give a number of examples to show how this obstruction can be calculated in favorable cases. Moreover, the classifying space $B(G;E)$ for E-oriented stable G-bundles ($G = O, U, F, Top$, etc.) is itself an infinite loop space, and the Thom spectrum $M(G;E)$ associated to the universal E-oriented GV-bundles, $V \subset R^\infty$, is again an E_∞ ring spectrum. Thus we have new cohomology theories of E-oriented G-bundles and corresponding new cobordism theories.

Our work raises many unanswered questions. Can these new theories be calculated in interesting cases ? What is the relationship between the cohomology theory determined by an E_∞ ring spectrum E and that determined by FE? What are the implications for $\pi_* E$ of the existence of an E_∞ ring structure on E? All that we show here is that the representing spectra of most of the interesting cohomology theories have an enormously richer internal structure than has yet been studied and that this structure gives rise to a collection of potentially powerful new theories.

1. E_∞ ring prespectra and spectra

One way to prove that a space is an infinite loop space is to display an action of an E_∞ operad on it [VI,§1]. We think of spectra as already carrying additive structures, and we shall superimpose multiplicative structures by means of actions by operads. Since spectra, as defined in II, 1.1, are indexed on the finite-dimensional sub inner product spaces of R^∞, it is natural to give the linear isometries operad $\mathcal{L}$ of I. 1.2 a privileged role in our theory. For the examples in this chapter, it would suffice to use only $\mathcal{L}$, but it is essential for later chapters to allow more general operads. Thus we assume given an E_∞ operad $\mathcal{G}$ and a morphism of operads $\mathcal{G} \to \mathcal{L}$. By abuse, we shall think of elements of $\mathcal{G}(j)$ as linear isometries via the given map to $\mathcal{L}(j)$.

Actions by operads refer to chosen basepoints. We think of actions by $\mathcal{G}$ on spaces as multiplicative, and the relevant basepoint is denoted by 1. We do not want to impose additive structures on spaces, but we do want to impose zeroes. Thus let $\mathcal{J}_e$ denote the category of spaces

X together with cofibrations $e: S^0 \to X$ where $S^0 = \{0,1\}$. By a $\mathcal{C}$-space with zero, or $\mathcal{C}_0$-space, we understand a $\mathcal{C}$-space (X,ξ) such that $X \in \mathcal{J}_e$ and $\xi_j(g, x_1, \ldots, x_j) = 0$ if any $x_i = 0$; in other words, $\xi_j: \mathcal{C}(j) \times X^j \to X$ is required to factor through the equivariant half-smash $\mathcal{C}(j) \times_{\Sigma_j} X \wedge \ldots \wedge X / \mathcal{C}(j) \times_{\Sigma_j} *$ defined with respect to the basepoint 0. Let $\mathcal{C}[\mathcal{J}_e]$ denote the category of $\mathcal{C}$-spaces with zero.

The spaces TV of a prespectrum have given basepoints, which we denote by 0; all wedges and smash products used below are to be taken with respect to these basepoints. By a unit for $T \in \mathcal{P}$ or for $E \in \mathcal{S}$ we understand a map $e: \Sigma^\infty S^0 \to T$ in $\mathcal{P}$ or $e: Q_\infty S^0 \to E$ in $\mathcal{S}$ such that the resulting (and, by II.1.3 and II.1.7, determining) map $S^0 \to T_0$ or $S^0 \to E_0$ is a cofibration. Let $\mathcal{P}_e$ and $\mathcal{S}_e$ denote the categories of prespectra with units and spectra with units (and morphisms which preserve units). The constructions and results of II §1 extend immediately to the categories $\mathcal{J}_e$, $\mathcal{P}_e$, and $\mathcal{S}_e$.

The reader is advised to review the definitions of operads (VI.1.2), of actions by operads (VI.1.3), of the linear isometries operad (I.1.2), and of prespectra and spectra (II.1.1) before proceeding to the following definition.

<u>Definition 1.1.</u> A $\mathcal{C}$-prespectrum (T, σ, ξ) is a unital prespectrum (T, σ) together with maps

$$\xi_j(g): TV_1 \wedge \ldots \wedge TV_j \to Tg(V_1 \oplus \ldots \oplus V_j)$$

for $j \geq 0$, $g \in \mathcal{C}(j)$, and $V_i \in \mathcal{J}_*(R^\infty)$, where $\xi_0(*)$ is to be interpreted as the inclusion $e: S^0 \to T_0$, such that the following conditions are satisfied.

(a) If $g \in \mathcal{C}(k)$, $h_r \in \mathcal{C}(j_r)$ for $1 \leq r \leq k$, and $j = j_1 + \ldots + j_k$, then the following diagram is commutative.

$$\begin{array}{ccc} TV_1 \wedge \ldots \wedge TV_j & \xrightarrow{\xi_j(\gamma(g;h_1,\ldots,h_k))} & T\gamma(g;h_1,\ldots,h_k)(V_1 \oplus \ldots \oplus V_j), \\ \Big\downarrow \xi_{j_1}(h_1) \ \ldots \ \xi_{j_k}(h_k) & & \Big\| \\ TW_1 \wedge \ldots \wedge TW_k & \xrightarrow{\xi_k(g)} & T g(W_1 \oplus \ldots \oplus W_k) \end{array}$$

where $W_r = h_r(V_{j_1+\ldots+j_{r-1}+1} \oplus \ldots \oplus V_{j_1+\ldots+j_r})$ (or $\{0\}$ if $j_r = 0$).

(b) $\xi_1(1)\colon TV \to TV$ is the identity map.

(c) If $g \in \mathcal{J}(j)$ and $\tau \in \Sigma_j$, then the following diagram is commutative:

$$\begin{array}{ccc} TV_1 \wedge \ldots \wedge TV_j & \xrightarrow{\xi_j(g\tau)} & Tg\tau(V_1 \oplus \ldots \oplus V_j) \\ \Big\downarrow \tau & & \Big\| \\ TV_{\tau^{-1}(1)} \wedge \ldots \wedge TV_{\tau^{-1}(j)} & \xrightarrow{\xi_j(g)} & Tg(V_{\tau^{-1}(1)} \oplus \ldots \oplus V_{\tau^{-1}(j)}) \end{array}$$

(d) For fixed V_i and W, ξ_j is continuous in g as g ranges through the subspace of $\mathcal{J}(j)$ which consists of those elements such that $g(V_1 \oplus \ldots \oplus V_j) = W$.

(e) If $g \in \mathcal{J}(j)$ and $V_i \perp W_i$, then the following diagram is commutative:

$$\begin{array}{ccc} TV_1 \wedge tW_1 \wedge \ldots \wedge TV_j \wedge tW_j & \xrightarrow{\sigma \wedge \ldots \wedge \sigma} & T(V_1 + W_1) \wedge \ldots \wedge T(V_j + W_j) \\ \Big\| \wr & & \Big\downarrow \xi_j(g) \\ TV_1 \wedge \ldots \wedge TV_j \wedge t(W_1 \oplus \ldots \oplus W_j) & & Tg((V_1 + W_1) \oplus \ldots \oplus (V_j + W_j)) \\ \Big\downarrow \xi_j(g) \wedge tg & & \Big\| \\ Tg(V_1 \oplus \ldots \oplus V_j) \wedge tg(W_1 \oplus \ldots \oplus W_j) & \xrightarrow{\sigma} & T(g(V_1 \oplus \ldots \oplus V_j) + g(W_1 \oplus \ldots \oplus W_j)) \end{array}$$

(where t is a sphere-valued functor on $\mathcal{J}_*$ as in II §1).

(f) If $g \in \mathcal{G}(1)$, then $\xi_1(g): TV \to TgV$ is a homeomorphism in the homotopy class $T(g|V)$, and every morphism $f \in \mathcal{J}_*(R^\infty)$ is obtainable by restriction from some $g \in \mathcal{G}(1)$. ((f) could be deleted; see Remarks II.1.10.)

(T,σ,ξ) is a $\mathcal{G}$-spectrum if (T,σ) is a spectrum. A morphism $\psi:(T,\sigma,\xi) \to (T',\sigma',\xi')$ of $\mathcal{G}$-prespectra is a morphism $\psi:(T,\sigma) \to (T',\sigma')$ of unital prespectra such that the following diagrams are commutative:

$$\begin{array}{ccc}
TV_1 \wedge \cdots \wedge TV_j & \xrightarrow{\xi_j(g)} & Tg(V_1 \oplus \cdots \oplus V_j) \\
\downarrow{\scriptstyle \psi\wedge\cdots\wedge\psi} & & \downarrow{\scriptstyle \psi} \\
T'V_1 \wedge \cdots \wedge T'V_j & \xrightarrow{\xi'_j(g)} & T'g(V_1 \oplus \cdots \oplus V_j)
\end{array}$$

Let $\mathcal{G}[\mathcal{P}_e]$ denote the category of $\mathcal{G}$-prespectra and let $\mathcal{G}[\mathcal{S}_e]$ denote its full subcategory of $\mathcal{G}$-spectra. Let $\nu: \mathcal{G}[\mathcal{S}_e] \to \mathcal{G}[\mathcal{P}_e]$ denote the inclusion functor.

<u>Definition 1.2.</u> An E_∞ ring prespectrum (or spectrum) is a $\mathcal{G}$-prespectrum (or $\mathcal{G}$-spectrum) over any E_∞ operad $\mathcal{G}$ with a given morphism of operads $\mathcal{G} \to \mathcal{L}$.

We have not defined and do not need any notion of a morphism between E_∞ ring spectra over different operads.

Think of a prespectrum (T,σ) as determining an underlying space, the wedge over all $V \in \mathcal{J}_*(R^\infty)$ of the spaces TV. Then conditions (a), (b), and (c) are precisely the algebraic identities required for the ξ_j to give this space a structure of $\mathcal{G}_0$-space. Condition (d) describes how to weave in the topology of $\mathcal{G}$, but we should add that we only know how to make effective use of the topology when the V_i and W are all $\{0\}$. The last two conditions relate the ξ_j to the internal structure of (T,σ). In practice, (f) is used to define the maps $T(g|V)$, and the force of the definition lies in condition (e).

In [48,XI], a smash product functor $\wedge_g: \mathcal{S} \times \mathcal{S} \to \mathcal{S}$ is defined for each element $g \in \mathcal{L}(2)$; all such functors become equivalent in the stable homotopy cagegory $H\mathcal{S}$. Our definition ensures that, for each $\mathcal{G}$-spectrum E and each $g \in \mathcal{G}(2)$, there is a well-defined map $E \wedge_g E \to E$ in $\mathcal{S}$ which gives E a structure of commutative ring spectrum in $H\mathcal{S}$.

Although irrelevant to our theory, a comparison with Whitehead's notion of a ring spectrum may be illuminating.

Remarks 1.3. Let (T, σ, ξ) be a $\mathcal{G}$-prespectrum and let $g \in \mathcal{G}(2)$. Let R^∞ and $R^\infty \oplus R^\infty$ have orthonormal bases $\{e_i\}$ and $\{e_i', e_i''\}$. Assume that tV is the one-point compactification of V and let $T_i = TR^i$ and $\sigma_i = \sigma: \Sigma T_i = T_i \wedge tRe_{i+1} \to T_{i+1}$. Consider the following diagram for any $p \geq 0$ and $q \geq 0$, where $d: Rge'_{p+1} \to Rge'_{q+1}$ is the obvious linear isometry and $f: g(R^p \oplus R^{q+1}) \to R^{p+q+1}$ is any linear isometry:

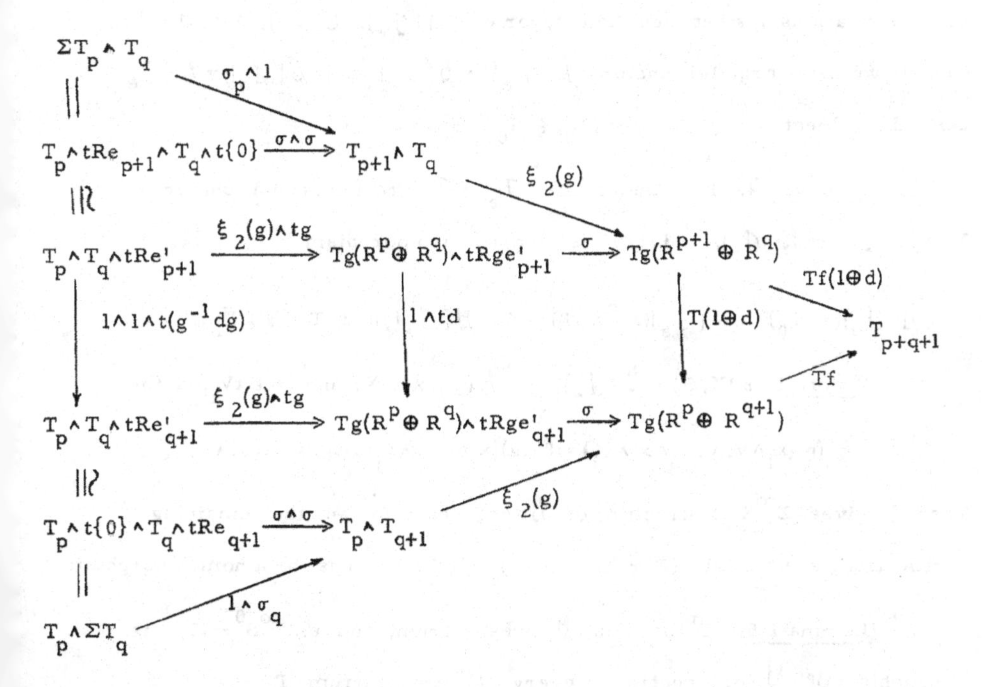

The left rectangle commutes trivially, the two trapezoids commute by (e) in the definition of a $\mathcal{H}$-prespectrum, and the two left triangles commute while the remaining rectangle and triangle homotopy commute by the very definition of a prespectrum. This looks just like Whitehead's diagram (see IX. 2.5), except that we haven't mentioned signs. The point is that, to get a pairing in Whitehead's sense, we must use fixed chosen isometries $f_{pq}: g(R^p \oplus R^q) \to R^{p+q}$. If $f = f_{p,q+1}$ in our diagram, then $f_{p+1,q}$ may lie in the opposite component from $f_{p,q+1}(1 \oplus d)$ in the space of linear isometries $g(R^{p+1} \oplus R^q) \to R^{p+q+1}$. Of course, our theory requires no such choices, and the linear isometries in the definition of a prespectrum efficiently keep track of all such changes of coordinates.

The interest lies in E_∞ ring spectra, but it is E_∞ spaces and ring prespectra which occur in nature. We next show that the relationships between the categories $\mathcal{J}_e$, $\mathcal{P}_e$, and $\mathcal{S}_e$ derived in II, §1 restrict to give similar relationships between the categories $\mathcal{H}[\mathcal{J}_e]$, $\mathcal{H}[\mathcal{P}_e]$, and $\mathcal{H}[\mathcal{S}_e]$. Clearly we have forgetful functors $\mathcal{H}[\mathcal{P}_e] \to \mathcal{H}[\mathcal{J}_e]$ and $\mathcal{H}[\mathcal{S}_e] \to \mathcal{H}[\mathcal{J}_e]$ defined on objects by $(T, \sigma, \xi) \to (T_0, \xi | T_0)$.

Lemma 1.4. The functor $\Sigma^\infty: \mathcal{J}_e \to \mathcal{P}_e$ restricts to a functor $\Sigma^\infty: \mathcal{H}[\mathcal{J}_e] \to \mathcal{H}[\mathcal{P}_e]$, and there is a natural isomorphism

$$\mathcal{H}[\mathcal{J}_e](X, T_0) \cong \mathcal{H}[\mathcal{P}_e](\Sigma^\infty X, T), \quad X \in \mathcal{H}[\mathcal{J}_e] \text{ and } T \in \mathcal{H}[\mathcal{P}_e].$$

Proof. For $(X, \xi) \in \mathcal{H}[\mathcal{J}_e]$, $g \in \mathcal{H}(j)$, $x_i \in X$, and $v_i \in tV_i$, define

$$\xi_j(g)(x_1 \wedge v_1 \wedge \ldots \wedge x_j \wedge v_j) = \xi_j(g)(x_1 \wedge \ldots \wedge x_j) \wedge (tg)(v_1 \wedge \ldots \wedge v_j).$$

Then ξ gives $\Sigma^\infty X$ a structure of $\mathcal{H}$-prespectrum, and the remaining verifications are trivial. (For (f), each $\xi_1(g): X \to X$ must be a homeomorphism.)

Lemma 1.5. $\Sigma^\infty S^0$ is a $\mathcal{H}$-prespectrum, and $e: \Sigma^\infty S^0 \to T$ is a morphism of $\mathcal{H}$-prespectra for every $\mathcal{H}$-prespectrum T.

Proof. With $\xi_j(g) = 1$ on $S^0 \wedge \ldots \wedge S^0 = S^0$ for each $g \in \mathcal{G}(j)$, S^0 is a $\mathcal{G}$-space with zero such that $e: S^0 \to X$ is a morphism of $\mathcal{G}$-spaces with zero for all $X \in \mathcal{G}[\mathcal{J}_e]$. The conclusion follows by taking $X = T_0$ and applying the previous lemma.

Lemma 1.6. The functor $\Omega^\infty: \mathcal{P}_e \to \mathcal{S}_e$ restricts to a functor $\Omega^\infty: \mathcal{G}[\mathcal{P}_e] \to \mathcal{G}[\mathcal{S}_e]$, and there is a natural isomorphism

$$\mathcal{G}[\mathcal{P}_e](T, \nu E) \cong \mathcal{G}[\mathcal{S}_e](\Omega^\infty T, E), \qquad T \in \mathcal{G}[\mathcal{P}_e] \text{ and } E \in \mathcal{G}[\mathcal{S}_e].$$

Proof. By II.1.4, $(\Omega^\infty T)(V) = \varinjlim \Omega^W T(V+W)$, $W \perp V$. Let $(T, \sigma, \xi) \in \mathcal{G}[\mathcal{P}_e]$, $g \in \mathcal{G}(j)$, and $f_i \in \Omega^{W_i} T(V_i + W_i)$ and define a map $\xi_j(g)(f_1 \wedge \ldots \wedge f_j)$ by commutativity of the diagram

$$\begin{array}{ccc}
tg(W_1 \oplus \ldots \oplus W_j) & \xrightarrow{\xi_j(g)(f_1 \wedge \ldots \wedge f_j)} & T(g(V_1 \oplus \ldots \oplus V_j) + g(W_1 \oplus \ldots \oplus W_j)) \\
\downarrow (tg)^{-1} & & \| \\
t(W_1 \oplus \ldots \oplus W_j) & & Tg((V_1 + W_1) \oplus \ldots \oplus (V_j + W_j)) \\
\| & & \uparrow \xi_j(g) \\
tW_1 \wedge \ldots \wedge tW_j & \xrightarrow{f_1 \wedge \ldots \wedge f_j} & T(V_1 + W_1) \wedge \ldots \wedge T(V_j + W_j)
\end{array}$$

$\xi_j(g): (\Omega^\infty T)(V_1) \wedge \ldots \wedge (\Omega^\infty T)(V_j) \to (\Omega^\infty T)g(V_1 \oplus \ldots \oplus V_j)$ is obtained by passage to limits, and these maps are easily verified to make $\Omega^\infty T$ a $\mathcal{G}$-spectrum. With $W_i = \{0\}$ in our diagram, we see that the map $\iota: T \to \nu\Omega^\infty T$ of II.1.5 is a morphism of $\mathcal{G}$-prespectra, and the rest is clear.

The previous two lemmas imply the following result.

Lemma 1.7. $S = \Omega^\infty \Sigma^\infty S^0$ is a $\mathcal{G}$-spectrum, and $e: S \to E$ is a morphism of $\mathcal{G}$-spectra for every $\mathcal{G}$-spectrum E.

$Q_\infty = \Omega^\infty \Sigma^\infty$, and Lemmas 1.4 and 1.6 can be composed.

Lemma 1.8. The functor $Q_\infty: \mathcal{J}_e \to \mathcal{S}_e$ restricts to a functor

$Q_\infty : \mathcal{H}[\mathcal{J}_e] \to \mathcal{H}[\mathcal{S}_e]$, and there is a natural isomorphism

$$\mathcal{H}[\mathcal{J}_e](X, E_0) \cong \mathcal{H}[\mathcal{S}_e](Q_\infty X, E), \qquad X \in \mathcal{H}[\mathcal{J}_e] \text{ and } E \in \mathcal{H}[\mathcal{S}_e].$$

The following immediate consequence of this lemma will be needed in VII. Recall that $QX = (Q_\infty X)_0$.

Lemma 1.9. The monad (Q, μ, η) in $\mathcal{J}_e$ restricts to a monad in $\mathcal{H}[\mathcal{J}_e]$; for $E \in \mathcal{H}[\mathcal{S}_e]$, the natural map $QE_0 \to E_0$ gives E_0 a structure of Q-algebra in $\mathcal{H}[\mathcal{J}_e]$.

As we shall see in VII, the lemma implies that E_0 is an "E_∞ ring space", which is a space with two E_∞ space structures so interrelated that the underlying H-space structures satisfy the distributivity laws up to all possible higher coherence homotopies. Moreover, we shall see that a connective $\mathcal{H}$-spectrum E can be reconstructed (up to homotopy) from the E_∞ ring space E_0.

Lemma 1.8 gives the following class of examples. Recall that any infinite loop space is an E_∞ space (VII.2.1) and any grouplike E_∞ space is an infinite loop space (VII.3.2).

Example 1.10. For any $\mathcal{H}$-space (X, ξ), without zero, construct a $\mathcal{H}$-space (X^+, ξ) with zero by adjoining a disjoint basepoint 0 to X and extending ξ in the evident way. $Q_\infty X^+$ is then a $\mathcal{H}$-spectrum, and the inclusion of X in QX^+ is a morphism of $\mathcal{H}$-spaces. If $\mathcal{H}'$ is any E_∞ operad and $\mathcal{H} = \mathcal{H}' \times \mathcal{L}$, then any $\mathcal{H}'$-space is a $\mathcal{H}$-space via the projection $\mathcal{H} \to \mathcal{H}'$, while the projection $\mathcal{H} \to \mathcal{L}$ allows $\mathcal{H}$ to be used in the present theory. Therefore $Q_\infty X^+$ is an E_∞ ring spectrum for any E_∞ space X.

2. $\mathcal{J}_*$-prefunctors and Thom spectra

As explained in I§1, to construct an $\mathcal{L}$-space it is often simplest to first construct an $\mathcal{J}_*$-functor. Analogously, to construct an $\mathcal{L}$-prespectrum, it is often simplest to first construct an $\mathcal{J}_*$-prefunctor.

<u>Definition 2.1.</u> An $\mathcal{J}_*$-prefunctor (T, ω, e) is a continuous functor $T: \mathcal{J}_* \to \mathcal{T}$ together with a commutative, associative, and continuous natural transformation $\omega: T \times T \to T \cdot \oplus$ (of functors $\mathcal{J}_* \times \mathcal{J}_* \to \mathcal{T}$) and a continuous natural transformation $e: t \to T$ such that

(a) $\omega: TV \times TW \to T(V \oplus W)$ factors through $TV \wedge TW$.

(b) The composite $TV \wedge tW \xrightarrow{1 \wedge e} TV \wedge TW \xrightarrow{\omega} T(V \oplus W)$ has adjoint an inclusion with closed image and coincides with the identity map of TV when $W = \{0\}$.

(c) The diagram
$$\begin{array}{ccc} t(V \oplus W) & \xrightarrow{e} & T(V \oplus W) \\ \| & & \uparrow \omega \\ tV \wedge tW & \xrightarrow{e \wedge e} & TV \wedge TW \end{array}$$
is commutative.

A morphism $\Phi: (T, \omega, e) \to (T', \omega', e')$ of $\mathcal{J}_*$-prefunctors is a continuous natural transformation $\Phi: T \to T'$ such that $\omega'(\Phi \times \Phi) = \Phi\omega$ and $e' = \Phi e$. The $\mathcal{J}_*$-functor t with $\omega: tV \times tW \to t(V \oplus W)$ the projection is also an $\mathcal{J}_*$-prefunctor with e the identity. Condition (c) asserts that $e: t \to T$ is a morphism of $\mathcal{J}_*$-prefunctors.

<u>Lemma 2.2.</u> An $\mathcal{J}_*$-prefunctor (T, ω, e) naturally determines an $\mathcal{L}$-prespectrum (T, σ, ξ).

<u>Proof.</u> The continuity of T ensures that its restriction to sub inner product spaces of R^∞ induces a functor $h\mathcal{J}_*(R^\infty) \to h\mathcal{T}$. Define $\sigma = \omega(1 \wedge e): TV \wedge tW \to T(V + W)$ for orthogonal pairs (V, W) of subspaces of R^∞. Then (a)-(c) and the associativity of ω ensure that (T, σ)

is a prespectrum and that $e: \Sigma^{\infty}S^0 \to T$ is a morphism of prespectra. For $V_i \subset R^{\infty}$ and $g \in \mathcal{L}(j)$, define

$$\xi_j(g): TV_1 \wedge \dots \wedge TV_j \to Tg(V_1 \oplus \dots \oplus V_j)$$

to be the composite

$$TV_1 \wedge \dots \wedge TV_j \xrightarrow{\omega} T(V_1 \oplus \dots \oplus V_j) \xrightarrow{T(g|V_1 \oplus \dots \oplus V_j)} Tg(V_1 \oplus \dots \oplus V_j).$$

It is straightforward to verify that (T, σ, ξ) is then an $\mathcal{L}$-prespectrum.

As pointed out to us by Becker, Kochman, and Schultz, there is a class of $\mathcal{J}_*$-functors which leads via Lemmas 2.2 and 1.6 to certain of the E_{∞} ring spectra $Q_{\infty}X^+$ of Example 1.10.

__Example 2.3.__ Let X be an Abelian topological monoid with product ϕ and unit η. Define an $\mathcal{J}_*$-prefunctor (TX, ω, e) by

$$(TX)(V) = X^+ \wedge tV \quad \text{and} \quad (TX)(f) = 1 \wedge tf \quad \text{for} \quad f: V \to V',$$

with ω and e given by the maps

$$\phi^+ \wedge 1: X^+ \wedge tV \wedge X^+ \wedge tW \cong (X \times X)^+ \wedge tV \wedge tW \to X^+ \wedge t(V \oplus W)$$

and

$$\eta^+ \wedge 1: tV \cong \{1\}^+ \wedge tV \to X^+ \wedge tV .$$

Note that t is recovered as the special case $T\{1\}$.

The $\mathcal{L}$-spectrum determined by t is $Q_{\infty}S^0$. The derived $\mathcal{L}$-space structure on the zeroth space QS^0 coincides with the $\mathcal{L}$-space structure derived from the $\mathcal{J}_*$-functor $\widetilde{F}$ defined in I.2.5. This phenomenon generalizes to arbitrary $\mathcal{J}_*$-prefunctors.

__Definition 2.4.__ Let (T, ω, e) be an $\mathcal{J}_*$-prefunctor and write M for the $\mathcal{L}$-spectrum derived by application of the functor Ω^{∞} to the associated $\mathcal{L}$-prespectrum T. Define an $\mathcal{J}_*$-functor $\widetilde{F}T$ by

$$\widetilde{F}TV = \Omega^V TV \quad \text{and} \quad \widetilde{F}Tf = \Omega^{f^{-1}} Tf \quad \text{for} \quad f: V \to V',$$

with $\omega: \widetilde{F}TV \times \widetilde{F}TW \to \widetilde{F}T(V \oplus W)$ given by the composite

$$\Omega^V TV \times \Omega^W TW \xrightarrow{\wedge} \Omega^{V\oplus W}(TV\wedge TW) \xrightarrow{\Omega^{V\oplus W}\omega} \Omega^{V\oplus W}T(V\oplus W).$$

As explained in I.1.6 and I.1.9, $\tilde{F}T$ determines an $\mathcal{L}$-space (also denoted by $\tilde{F}T$) by passage to limits over $V \subset R^\infty$. A trivial comparison of definitions shows that $\tilde{F}T = M_0$ as an $\mathcal{L}$-space. There are evident sub $\mathcal{J}_*$-functors FT and SFT of $\tilde{F}T$ which give rise to the sub $\mathcal{L}$-spaces FM and SFM of M_0.

We next display the Thom spectra as $\mathcal{J}_*$-prefunctors. Recall the discussion of the two-sided geometric bar construction as an $\mathcal{J}_*$-functor from I.2.1 and I.2.2.

Construction 2.5. Let $G \to \tilde{F}$ be a morphism of monoid-valued $\mathcal{J}_*$-functors. Then G acts from the left by evaluation on the $\mathcal{J}_*$-functor t and from the left and right on the trivial $\mathcal{J}_*$-functor $*$. Let Y be any $\mathcal{J}_*$-functor on which G acts from the right. The map $p: B(YV, GV, tV) \to B(YV, GV, *)$ is a quasi-fibration if G is grouplike and a GV-bundle if G is group-valued. p admits a cross-section σ induced from the morphism of GV-spaces $* \to tV$ and has fibre $\tau: tV \to B(YV, GV, tV)$ over the basepoint of $B(YV, GV, *)$. Moreover, p, σ, and τ are all morphisms of $\mathcal{J}_*$-functors. Define an $\mathcal{J}_*$-prefunctor $(T(G;Y), \omega, e)$ by

$$T(G;Y)(V) = B(YV, GV, tV)/B(YV, GV, *),$$

with ω and e induced from the Whitney sum of $B(Y, G, t)$ and from τ. Write $T(G;Y)$ for the associated $\mathcal{L}$-prespectrum and write $M(G;Y)$ for the derived $\mathcal{L}$-spectrum $\Omega^\infty T(G;Y)$. In an evident sense, T and M are functorial on pairs (G, Y). Abbreviate $T(G;*) = TG$ and $M(G;*) = MG$. TG and MG are called the Thom prespectrum and spectrum of G.

Clearly Me coincides with $Q_\infty S^0$. When G is group-valued, we could just as well define MG by use of the associated sphere bundles to the principal bundles

$$G(V \oplus V)/e \times GV \rightarrow G(V \oplus V)/GV \times GV.$$

However, since these bundles are not universal (because their total spaces are not contractible), it seems preferable even in the classical case to use the bar construction.

When G = F, p must be replaced by an appropriate fibration DFV → BFV in order to obtain a universal FV-bundle (spherical fibration with cross-section). Here we could replace TFV by DFV/BFV; the new TFV would again determine an $\mathcal{J}_*$-prefunctor and the old TFV would be deformation retracts of the new ones via deformations which define morphisms of $\mathcal{J}_*$-prefunctors for each parameter value. A similar remark applies to the general case T(G; Y) when G maps to F.

In view of I. 2. 5, all of the usual cobordism theories except for PL theory are thus represented by $\mathcal{L}$-spectra. While it may be possible to handle MPL and MSPL in an ad hoc manner based on the triangulation theorem, as BPL and BSPL were handled in I §2, it is certainly preferable to treat these within a general framework of axiomatic bundle theory. Such a treatment will be given by the second author in [64].

We note one other important example to which our theory does not yet apply, namely the Brown-Peterson spectrum. The point is that our theory requires a good concrete geometric model, not merely a homotopy theoretical construction, and no such model is presently known for BP.

For general Y in Construction 2. 5, Lashof's treatment [36] of the Pontryagin-Thom construction implies that if G is group-valued and maps to O, then $\pi_* M(G; Y)$ gives the cobordism groups of G-manifolds with a

"Y-structure" on their stable normal bundles. In full generality, when G maps to F, define a G-normal space to be a normal space in the sense of [63,1.1] with a reduction of the structural monoid of its spherical fibration to G [47,10.4]; then $\pi_* M(G;Y)$ gives the cobordism groups of G-normal spaces with a "Y-structure" on their spherical fibrations. An intrinsic bundle or fibration theoretic interpretation of a Y-structure requires an appropriate classification theorem, and general results of this nature are given in [47,§11]. When G maps to G', a G'-structure is a G'-trivialization [47,10.3]. When H maps to G, an H\G-structure is a reduction of the structural monoid to H [47,10.4]. When Y = FM is as in Definition 2.4 and is regarded as a right $\mathcal{J}_*$-functor over G via composition of maps,

$$FMV \times GV \rightarrow F(tV, TV) \times F(tV, tV) \rightarrow F(tV, TV) = FMV,$$

a Y-structure is an M-orientation by III § 2 and Remarks 3.5 below.

Note that the map of $\mathcal{L}$-prespectra $\iota : TG \rightarrow \nu\Omega^{\infty}TG = \nu MG$ defines explicit MG-orientations $\iota : TGV \rightarrow MGV$ of the universal GV-bundle (via the equivalence of TFV and DFV/BFV when G = F). Thus any GV-bundle admits an MG-orientation. The following lemma reflects the fact that a G-normal space (or G-manifold) with an H\G-structure admits an FMH-structure (that is, an MH-orientation).

<u>Lemma 2.6.</u> Let $H \rightarrow G \rightarrow \tilde{F}$ be morphisms of monoid-valued $\mathcal{J}_*$-functors. Then there is a morphism $j: H\backslash G \rightarrow \tilde{F}MH$ of right $\mathcal{J}_*$-functors over G such that j coincides with the given morphism $G \rightarrow \tilde{F}$ when H = e and the following diagram commutes:

$$\begin{array}{ccc} G & \xrightarrow{\tau} & H\backslash G \\ \downarrow j & & \downarrow j \\ \tilde{F} & \xrightarrow{e} & \tilde{F}MH \end{array}$$

If G maps to F or SF then H\G maps to FMH or SFMH.

Proof. $H\backslash G = B(*, H, G)$ and we have the commutative diagram

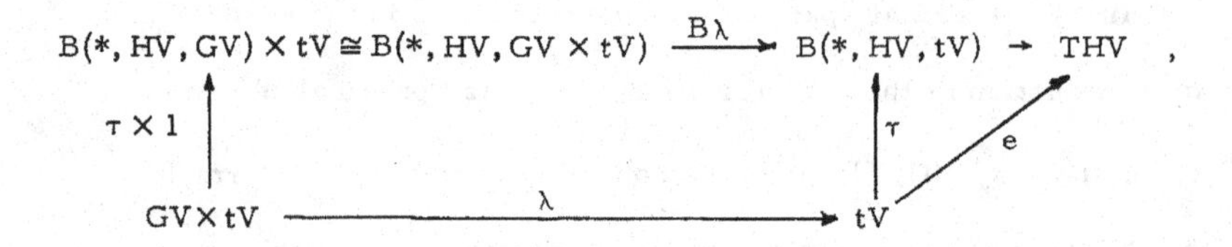

where λ is the evaluation map. $j: HV\backslash GV \to F(tV, THV) = \widetilde{F}MH$ is defined to be the adjoint of the top composite.

As will be discussed and interpreted geometrically in [66], the maps $j: H\backslash G \to FMH$ induce the bordism J-homomorphisms.

We record a number of natural maps of $\mathcal{J}_*$-prefunctors in the following remarks. The same letters will be used for the derived morphisms of $\mathcal{L}$-prespectra and $\mathcal{L}$-spectra. The cobordism interpretations should be clear from the discussion above.

Remarks 2.7 (i) For (G, Y) as in Construction 2.5, the morphism of $\mathcal{J}_*$-functors $q: B(Y, G, t) \to B(*, G, t)$ induces a morphism of $\mathcal{J}_*$-prefunctors $q: T(G; Y) \to TG$.

(ii) In the notation of the previous proof, the maps $B\lambda$ induce a morphism of $\mathcal{J}_*$-functors $\varepsilon = \varepsilon(B\lambda): B(H\backslash G, G, t) \to B(*, H, t)$ which in turn induces a morphism of $\mathcal{J}_*$-prefunctors $\varepsilon : T(G; H\backslash G) \to TH$.

(iii) If G maps to F and if T is an $\mathcal{J}_*$-prefunctor, then the evaluation maps $F(tV, TV) \times tV \to TV$ induce maps $\varepsilon: B(FMV, GV, tV) \to TV$ which in turn induce a morphism $\varepsilon: T(G, FM) \to T$ of $\mathcal{J}_*$-prefunctors.

The maps of the previous lemma and remarks give considerable information about the structure of $M(G; MG)$.

Remarks 2.8. For an $\mathcal{L}$-spectrum M derived from an $\mathcal{J}_*$-prefunctor, write

$$M(G; M) = \Omega^{\infty}T(G; FM) \quad \text{and} \quad M(SG; M) = \Omega^{\infty}T(SG; SFM).$$

Let $H \to G \to F$ be morphisms of monoid-valued $\mathcal{J}_*$-functors. The following diagrams commute because they already do so on the level of $\mathcal{J}_*$-prefunctors:

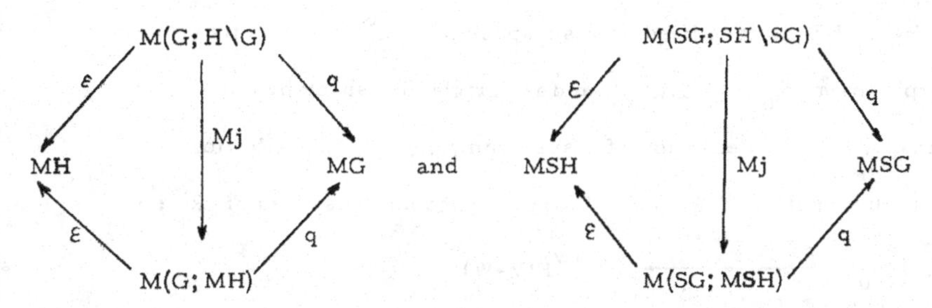

When $H = G$, $H\backslash G = B(*, G, G)$ is contractible and the upper maps ε and q are isomorphisms in $H\mathcal{S}$ (because the maps ε and q on the prespectrum level are weak homotopy equivalences for each V). We conclude that, in $H\mathcal{S}$, the lower maps q split off a direct factor MG or MSG (via $Mj \cdot \varepsilon^{-1}$) such that the restriction to this factor of the lower map ε is the identity.

3. Orientation theory for E_∞ ring spectra

Let $\mathcal{H}$ be an E_∞ operad with a given map to $\mathcal{L}$ and consider $\mathcal{L}$-spaces as $\mathcal{H}$-spaces via this map. We shall write BX for the first de-looping of a grouplike $\mathcal{H}$-space X (VII,§3 or [46]). This is a harmless abuse of notation since BX is equivalent as an infinite loop space to the usual classifying space of X if X happens to be a topological monoid in the category of $\mathcal{H}$-spaces (by VII, 3.6).

Let E be a $\mathcal{H}$-spectrum. By Lemma 1.7, $e: S \to E$ is a morphism of $\mathcal{H}$-spectra and thus $e: F \to FE$ is a morphism of $\mathcal{H}$-spaces. Let R denote the commutative ring $\pi_0 E$.

Let $j: G \to F$ be a morphism of monoid-valued $\mathcal{J}_*$-functors, and let j also denote the derived map of $\mathcal{H}$-spaces (I.1.9 and I.1.6).

Write e for any of the composites $GV \to G \xrightarrow{j} F \xrightarrow{e} FE$, $V \subset R^\infty$. Consider the orientation diagram constructed in III §2. It features the classifying space $B(GV;E) = B(FE, GV, *)$ for E-oriented GV-bundles. Here FE is identified as a subspace of $\Omega^V EV$ via the homeomorphism $\tilde{\sigma}: E_0 \to \Omega^V EV$ and is a right GV-space by composition of maps. By the definition of a spectrum, II, 1.1, the following diagram is commutative if V and W are orthogonal subspaces of R^∞:

$$\begin{array}{ccc} E_0 & \xrightarrow{\tilde{\sigma}} & \Omega^{V+W}E(V+W) \\ {\scriptstyle\tilde{\sigma}}\downarrow & & \Vert\wr \\ \Omega^V EV & \xrightarrow{\Omega^V\tilde{\sigma}} & \Omega^V\Omega^W E(V+W) \end{array}$$

Therefore the identification of FE as a subspace of $\Omega^V EV$ is consistent as V varies, and FE inherits a right action by $G = GR^\infty$ from the right actions by the GV. Moreover, the action $FE \times G \to FE$ is itself a morphism of $\mathcal{H}$-spaces. Indeed, a comparison of (e) of Definition 1.1 with I.1.10 shows that this follows from the cancellation $(xg^{-1})(gy) = xy$ for $g: V \to V'$, $x \in \Omega^V EV$, and $y \in \Omega^V tV = FV$. Now recall the discussion of the two-sided geometric bar construction as a $\mathcal{H}$-space from I §2 (or [46,§3]). Recall too that an Abelian monoid is a $\mathcal{H}$-space for any $\mathcal{H}$ and that the discretization $d: X \to \pi_0 X$ of a $\mathcal{H}$-space is always a map of $\mathcal{H}$-spaces [45,§3]. By passage to limits from the orientation diagrams for $V \subset R^\infty$, we conclude the following result.

<u>Theorem 3.1.</u> All spaces are grouplike $\mathcal{H}$-spaces and all maps are $\mathcal{H}$-maps in the stable orientation diagram

$$\begin{array}{ccccccc} SG & \xrightarrow{e} & SFE & \xrightarrow{\tau} & B(SG;E) & \xrightarrow{q} & BSG \\ {\scriptstyle i}\downarrow & & \downarrow{\scriptstyle i} & & \downarrow{\scriptstyle i} & & \downarrow{\scriptstyle i} \\ G & \xrightarrow{e} & FE & \xrightarrow{\tau} & B(G;E) & \xrightarrow{q} & BG \\ \Vert & & \downarrow{\scriptstyle d} & & \downarrow{\scriptstyle Bd} & & \Vert \\ G & \xrightarrow{de} & FR & \xrightarrow{\tau} & B(G;R) & \xrightarrow{q} & BG \end{array}$$

This diagram is therefore equivalent to the diagram obtained by application of the zeroth space functor to a diagram of connective spectra in which the rows are fiberings (that is, are equivalent in the stable category to fibration sequences [48, XI]). On the level of spaces, the stable E-orientation sequence now extends infinitely in both directions:

$$\cdots \to \Omega B(G;E) \to G \xrightarrow{e} FE \xrightarrow{\tau} B(G;E) \xrightarrow{q} BG \xrightarrow{Be} BFE \to BB(G;E) \to \cdots$$

Given $H \to G$, we also have the infinite sequence

$$\cdots \to \Omega(G/H) \to H \to G \xrightarrow{\tau} G/H \xrightarrow{q} BH \to BG \to B(G/H) \to \cdots$$

and a map of this sequence into the E-orientation sequence of H (because the maps of III.2.6 are $\mathcal{L}$-maps when $V = R^{\infty}$).

Since $B(G;E)$ is equivalent to the fibre of $Be: BG \to BFE$, an easy diagram chase shows that $B(GV;E)$ is equivalent to the fibre of the following composite, which we again denote by Be:

$$BGV \to BG \xrightarrow{Bj} BF \xrightarrow{Be} BFE.$$

<u>Definition 3.2.</u> Let ξ be a GV-bundle classified by $\alpha: X \to BGV$. Define $w(\xi;E)$ to be the element $\alpha^*(Be)$ of the group $[X^+, BFE]$. $w(\xi;E)$ is called the E-theory Stiefel-Whitney class of ξ and is the obstruction to its E-orientability. If FE also denotes the (reduced) cohomology theory represented by the spectrum determined by FE, then $w(\xi;E)$ can be regarded as an element of $FE^1(X^+)$.

Previously, the obstruction to E-orientability was studied by analysis of the Atiyah-Hirzebruch spectral sequence. Larry Taylor [74] has given a number of results so obtainable, and we are indebted to him for several very helpful conversations. When E is an E_∞ ring spectrum, these results are immediate consequences of the definition and ordinary obstruction theory. To see this, assume that X is a finite dimensional CW-complex. If $w(\xi;E):X \to BFE$ is null homotopic on the (n-1)-skeleton X^{n-1}, then we have an obstruction set

$$w_n(\xi;E) \subset H^n(X;\pi_n BFE) = H^n(X;\pi_{n-1}FE)$$

to the existence of a null homotopy of $w(\xi;E)$ on X^n. Clearly $w_1(\xi;E)$ is the usual Stiefel-Whitney obstruction to the R-orientability of ξ of III. 2.2. For $n > 0$,

$$\pi_n FE = \pi_n SFE = \pi_n(E_0, 1) \cong \pi_n(E_0, 0) = \pi_n E,$$

where the isomorphism is given by translation from the 1-component to the 0-component. Let ψ be a GW-bundle over Y classified by $\beta: Y \to BGW$. Then $\xi \wedge \psi$ is classified by the composite of $\alpha \times \beta$ and $\omega: BGV \times BGW \to BG(V \oplus W)$. Take V and W to be orthogonal subspaces of R^∞. Then the diagram

$$\begin{array}{ccc} BGV \times BGW & \xrightarrow{\omega} & BG(V+W) \\ \downarrow & & \downarrow \\ BG \times BG & \xrightarrow{\phi} & BG \end{array}$$

is homotopy commutative, where ϕ is the product given by the $\mathcal{J}$-space structure. Since $Be: BG \to BFE$ is an H-map, we conclude that $w_n(\xi \wedge \psi;E)$ is defined and contains $w_n(\xi;E) + w_n(\psi;E)$ if $w_n(\xi;E)$ and $w_n(\psi;E)$ are defined. Clearly w_n is natural in X, in the sense that if $f: X' \to X$ is a map and if $w_n(\xi;E)$ is defined, then $w_n(f^*\xi;E)$ is defined and contains $f^*w_n(\xi;E)$. Similarly, if $\theta: E \to E'$ is a morphism of E_∞ ring spectra, then $w_n(\xi;E')$ contains $\theta_* w_n(\xi;E)$. Of course, the E-orientability of ξ implies its E'-orientability under the much weaker assumption that θ is a morphism of ring spectra in $H\mathcal{S}$.

Since $\pi_* BF$ is finite in each degree and X is finite dimensional, $a(Bj\circ\alpha) = 0$ for some positive integer a. Therefore $aw_n(\xi;E) \subset w_n(a\xi;E) = 0$ if $w_n(\xi;E)$ is defined. Thus, if ξ is R-orientable and if $H^n(X;\pi_{n-1}E)$ is torsion free for $n > 1$, then ξ is E-orientable. For example, if $E = MU$ or if E is a ring spectrum into which MU maps (such as MO, MSO, KU, etc.) and if $H^{2n+1}(X;Z)$ is torsion free for $n \geq 1$, then ξ is E-orientable if it is R-orientable.

When X is finite, $w(\xi;E)$ will be null homotopic if and only if its localizations at all primes, or at and away from a set T of primes, are null homotopic. Let E_T and $E[T^{-1}]$ denote the localization of E at and away from T. Suppose that $G = O$. Then 4ξ admits a symplectic structure, hence is MSp-orientable. If $w_1\xi = 0$, it follows that ξ is MSp[1/2]-orientable and thus that ξ is MSp-orientable if and only if it is $(MSp)_2$-orientable. Since the same statement holds for ring spectra into which MSp maps and these include most of the interesting Thom spectra MG and KO and KU, E-orientability of vector bundles is generally only a problem at the prime 2.

By the definition of $w(\xi;E)$ and the third author's result [65] that $(ej)_*: \pi_n O \to \pi_n MSp$ is zero for $n \geq 2$, any vector bundle over S^n, $n \geq 3$, is E-orientable for any ring spectrum E into which MSp maps.

Clearly applications like this can be multiplied ad infinitum, and our context gives a conceptually satisfactory and computationally efficient framework for the analysis of E-orientability.

Returning to the stable orientation diagram, we note that if $E = M$ happens to be derived from an $\mathcal{J}_*$-prefunctor, then, by I, §2 and use of the $\mathcal{J}_*$-functor FM of Definition 2.4, this diagram is derived by passage to $\mathcal{L}$-spaces from a commutative diagram of $\mathcal{J}_*$-

functors. Analogously, although we cannot construct an $\mathcal{J}_*$-prefunctor like $T(G;FM)$ for a general $\mathcal{H}$-spectrum E, we can construct a Thom prespectrum $T(G;E)$ by direct appeal to Definition 1.1.

Construction 3.3. Define a $\mathcal{H}$-spectrum $T(G;E)$ as follows.

$$T(G;E)(V) = B(FE, GV, tV)/B(FE, GV, *)$$

and, for $f: V \to V'$, $T(G;E)(f)$ is induced from $B(1, Gf, tf)$. $e: tV \to T(G;E)(V)$ is induced from $\tau: tV \to B(FE, GV, tV)$. For V orthogonal to W, $\sigma: T(G;E)(V) \wedge tW \to T(G;E)(V+W)$ is induced from the composite

$$B(FE, GV, tV) \times tW \cong B(FE, GV, tV \times tW) \xrightarrow{B(1,i,\omega)} B(FE, G(V+W), t(V+W)),$$

where $i: GV \to G(V+W)$ is the natural inclusion. With these maps, $T(G;E)$ is a unital prespectrum. For $g \in \mathcal{H}(j)$, the maps

$$\xi_j(g): T(G;E)(V_1) \wedge \dots \wedge T(G;E)(V_j) \to T(G;E)(g(V_1 \oplus \dots \oplus V_j))$$

are induced from the composites

$$\begin{array}{c}
B(FE, GV_1, tV_1) \times \dots \times B(FE, GV_j, tV_j) \\
\Vert\wr \\
B((FE)^j, GV_1 \times \dots \times GV_j, tV_1 \times \dots \times tV_j) \\
\Big\downarrow B(\xi_j(g), Gg \circ \omega, tg \circ \omega) \\
B(FE, Gg(V_1 \oplus \dots \oplus V_j), tg(V_1 \oplus \dots \oplus V_j)) .
\end{array}$$

The verification that these maps are well-defined and give $T(G;E)$ a structure of $\mathcal{H}$-prespectrum is tedious, but quite straightforward. Write $T(SG;E)$ for the $\mathcal{H}$-prespectrum defined similarly but with G and FE replaced by SG and SFE. Write $M(G;E)$ and $M(SG;E)$ for the $\mathcal{H}$-spectra derived by application of the functor Ω^∞.

By [36], [63], and III§2, $\pi_* M(G;E)$ gives the cobordism groups of (normally) E-oriented G-manifolds when G maps to O and of

E-oriented G-normal spaces in general. In the following remarks we record certain morphisms of $\mathcal{H}$-spectra, show the consistency of our two definitions of M(G; E) when E is derived from an $\mathcal{J}_*$-prefunctor, and discuss the structure of M(G; E) when MG maps to E.

Remarks 3.4 (i). The maps $q: B(FE, GV, tV) \to B(*, GV, tV)$ induce a morphism $q: T(G; E) \to TG$ of $\mathcal{H}$-prespectra and thus a morphism $q: M(G; E) \to MG$ of $\mathcal{H}$-spectra.

(ii) If H maps to G, the maps $Be: B(GV, HV, tV) \to B(FE, HV, tV)$ induce a morphism $Te: T(H; G) \to T(H; E)$ of $\mathcal{H}$-prespectra and thus a morphism $Me: M(H; G) \to M(H; E)$ of $\mathcal{H}$-spectra.

(iii) The evaluation maps $FE \times tV \xrightarrow{\tilde{\sigma} \times 1} F(tV, EV) \times tV \to EV$ induce maps $\varepsilon: B(FE, GV, tV) \to EV$ which in turn induce a morphism $\varepsilon: T(G; E) \to \nu E$ of $\mathcal{H}$-prespectra and thus, by Lemma 1.6, a morphism $\varepsilon: M(G; E) \to E$ of $\mathcal{H}$-spectra.

Remarks 3.5. Let E = M be derived from an $\mathcal{J}_*$-prefunctor T. Then $M_0 = \varinjlim \Omega^V TV$ and the natural maps $\Omega^V TV \to M_0$ induce maps $B(FMV, GV, tV) \to B(FM, GV, tV)$ which in turn induce a morphism of $\mathcal{H}$-prespectra from T(G; FM) of Construction 2.5 and Lemma 2.2 to T(G; M) of Construction 3.3. In view of the limits used in the definition of Ω^∞, the induced map of $\mathcal{H}$-spectra is an identification. The maps ε and q on M(G; M) given in Remarks 2.7 and the previous remarks coincide.

Remarks 3.6. Let $\gamma: MG \to E$ be a morphism of $\mathfrak{U}$-spectra.

Then the following diagram is commutative:

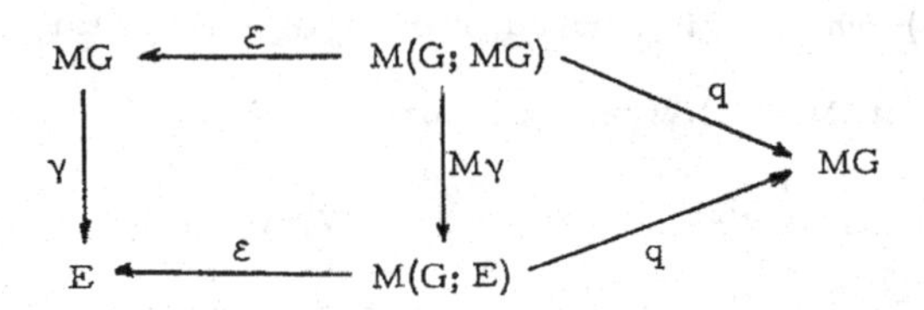

We conclude from Remarks 2.8 that, in $H\mathcal{S}$, $q: M(G; E) \to MG$ splits off a direct factor MG (via $M\gamma \circ Mj \circ \varepsilon^{-1}$) such that $\varepsilon: M(G; E) \to E$ restricts on this factor to the given map γ. We have an analogous result with G replaced by SG.

V. On kO-oriented bundle theories

One purpose of this chapter is to lay the foundations for an analysis of Adams' study of the groups $J(X)$ and Sullivan's study of topological bundle theory from the point of view of infinite loop space theory. For this purpose, it is essential to understand which portions of their work depend on the geometry (and representation theory) and which portions follow by purely formal manipulations on the classifying space level. It turns out that substantial parts of their results can be obtained by elementary chases of a pair of large diagrams focusing on the classifying space $B(SF;kO)$ for kO-oriented spherical fibrations and on BTop. The functors and natural transformations represented on finite-dimensional CW-complexes by the spaces and maps in these diagrams are easily described, and it is simple to interpret the information obtained on the classifying space level in bundle-theoretic terms.

For the construction and analysis of both diagrams, we shall take the following data as given. (More precise formulations of the data will be given later.)

(1) The Adams operations ψ^r and their values on $KO(S^n)$ [1].

(2) The validity of the Adams conjecture [2,17,57,73].

(3) The splitting of BO when localized at an odd prime [6,53].

For the first diagram, we shall also take as given

(4) The Atiyah-Bott-Shapiro kO-orientation of Spin bundles and the values of the derived cannibalistic classes ρ^r on $KSpin[1/r](S^n)$ [13,3].

Note that these results do not depend on Adams' last two $J(X)$ papers [4 and 5].

For the second diagram, we shall also take as given

(5) The Sullivan kO[1/2]-orientation of STop bundles, the fact that the induced map $F/Top[1/2] \to BO_{\otimes}[1/2]$ is an equivalence, and the values of the derived cannibalistic classes θ^r on $KO[1/2, 1/r](S^n)$ [71,72].

After stabilizing the classification theory for oriented bundles and fibrations developed in [47], we explain what we mean by an orientation of a stable bundle theory with respect to a cohomology theory in section 1. We construct certain general diagrams which relate oriented bundle theories to cohomology operations and to larger bundle theories in section 2.

We construct our first main diagram in section 3. By chasing its localizations, we derive splittings at each prime p of various spaces in the diagram, such as B(SF; kO), SF/Spin, and SF, in section 4. These splittings, and chases, imply many of Adams' calculations in [5]. The splittings of SF and F/O were noted by Sullivan (unpublished), but the recognition of the role played by B(SF;kO) and, following from this, the recognition that the analysis at the prime 2 is formally identical to that at the odd primes appear to be new.

In section 5, we prove a version of the main theorem of [4] and so recalculate the groups J(X). We also introduce bundle theoretic analogs δ and ε of the d and e invariants studied by Adams in [5]. δ gives the obstruction to kO-orientability of stable spherical fibrations (and its form depends on application of IV§3 to the E_∞ ring spectrum kO). ε is defined on the group Q(SF;kO)(X) of kO-orientable stable spherical fibrations over X and takes values in a certain group $JSpin_{\otimes}(X)$. Its restriction to JSpin(X) is an isomorphism. Therefore JSpin(X) is a direct summand of Q(SF;kO)(X); the complementary summand is the group of j-oriented stable spherical fibrations for a certain spectrum j. This analysis should be regarded as a generalization of that carried out by Adams [5] for the case when X is an i-sphere with $i > 2$.

In section 6, we construct our second main diagram. It looks just like the first one, and its analysis is exactly the same; only the interpretation changes. Chases of its localizations give Sullivan's splittings of BTop and Top/O at odd primes. These splittings, and chases, imply the odd primary part of Brumfiel's calculations [15]. Away from the prime 2, Q(SF; kO)(X) is isomorphic to JTop(X), the δ invariant becomes the obstruction to the existence of a Top-structure on a stable spherical fibration, and the ε-invariant yields Sullivan's analysis of JTop(X).

Very little of this theory depends on the use of infinite loop spaces. However, the machinery developed in this book shows that all spaces in sight are infinite loop spaces. This extra structure is essential to the applications. Characteristic classes for spherical fibrations, kO-oriented spherical fibrations, and topological bundles can only be described, at present, in terms of homology operations, and these operations are invariants of the infinite loop space structure. Thus it is important to know which of the splittings described here are only homotopical and which are as of infinite loop spaces. The problem, then, is to determine which of the maps displayed in our diagrams are infinite loop maps and which parts of the diagrams commute on the infinite loop space level. In section 7, by combining results of this book with recent results of Adams and Priddy [8] and still more recent results of Madsen, Snaith, and Tornehave [42] and Ligaard [38], we shall nearly complete the infinite loop analysis of our diagrams.

1. E-orientations of stable bundle theories

For the reader's convenience, we quickly summarize those notations from III to be used in this chapter. We then establish notations for stable bundle theories and explain what we mean by an orientation of such a theory with respect to a commutative ring spectrum E.

Let $j: G \to F$ be a morphism of grouplike monoid-valued $\mathcal{J}_*$-functors (I.1.8, 2.1, and 2.5) and let V and W be orthogonal finite-dimensional sub-inner product spaces of R^∞. BGV and B(GV; E) classify GV-bundles and E-oriented GV-bundles over CW-complexes, and $\omega: BGV \times BGW \to BG(V+W)$ and $B(\omega, \phi): B(GV; E) \times B(GW; E) \to B(G(V+W); E)$ induce the (external) fibre-wise smash product and (internal) Whitney sum (III §1, 2.1, and 2.8). There is an explicit quasi-fibration sequence

$$GV \xrightarrow{e} FE \xrightarrow{\tau} B(GV; E) \xrightarrow{q} BGV ,$$

the bundle-theoretic interpretation of which is given in III.2.5. If $i: H \to G$ is a morphism of grouplike monoid-valued $\mathcal{J}_*$-functors, there is another explicit quasi-fibration sequence

$$GV \xrightarrow{\tau} GV/HV \xrightarrow{q} BHV \xrightarrow{Bi} BGV$$

and a map $Be: GV/HV \to B(HV; E)$, interpreted in III.2.6, such that $qBe = q$. The maps τ and q of the two quasi-fibration sequences above are defined in the same way in terms of the bar construction and have analogous interpretations in terms of transformations to and from bundles with additional structure, hence the duplicative notation. If $\zeta: E \to E'$ is a map of ring spectra, there is a map $B\zeta: B(GV; E) \to B(GV; E')$, interpreted in III.2.9, such that $qB\zeta = q$. We write SG instead of G when all bundles are given with a canonical integral orientation, and then all E-orientations are required to be consistent with the preassigned integral orientation.

Write $G, G/H, BG$, and $B(G; E)$ for the spaces obtained by passage to limits over $V \subset R^\infty$. The first three are infinite loop spaces (by I) and the last is at least a grouplike H-space (by III.2.8) and is an infinite loop space if E is an E_∞ ring spectrum (by IV.3.1).

Henceforward, restrict attention to connected finite-dimensional CW-complexes X as base spaces of GV-bundles. GV and GV' bundles ξ and ξ' over X are said to be stably equivalent if $\xi \oplus \varepsilon W$ is equivalent to $\xi' \oplus \varepsilon W'$ for some $W \perp V$ and $W' \perp V'$ such that $V + W = V' + W'$. Write $\{\xi\}$ for the stable equivalence class of ξ and call $\{\xi\}$ a stable G-bundle. Let $\widetilde{K}G(X)$ denote the set of stable G-bundles over X. Then $\widetilde{K}G(X)$ is classified by BG, and the image of $[X^+, BGV]$ in $[X^+, BG]$ depends only on the dimension of V. The product on BG induces the external and internal operations $\wedge$ and $\oplus$ on stable G-bundles described in terns of the fibrewise smash product and Whitney sum by $\{\xi\} \wedge \{\psi\} = \{\xi \wedge \psi\}$ and $\{\xi\} \oplus \{\psi\} = \{\xi \oplus \psi\}$, where ξ and ψ are representative GV and GW bundles with $V \perp W$.

Let $KG(X)$ denote the Grothendieck group constructed from the $G(n)$-bundles for $n \geq 0$ (or, with the additional relations $\varepsilon V = \varepsilon V'$ when $\dim V = \dim V'$, from the GV-bundles for $V \subset R^\infty$). We have identifications

$$\widetilde{K}G(X) = [X^+, BG] = [X^+, BG \times \{0\}] \subset [X^+, BG \times Z] = KG(X).$$

Let $JG(X)$ denote the image of $\widetilde{K}G(X)$ in $\widetilde{K}F(X)$ under neglect of G-reduction (or, equivalently if G is group-valued, passage to fibre homotopy equivalence). Thus

$$JG(X) = (Bj)_*[X^+, BG] \subset [X^+, BF], \qquad Bj: BG \to BF.$$

Adams writes $\widetilde{J}(X)$ for $JO(X)$ and $J(X)$ for $JO(X) \times Z \subset KF(X)$; his notation is more logical, but less convenient since $JO(X) \times Z$ has no geometric ring structure and is therefore uninteresting. Of course, $J = j_*: \pi_*SO \to \pi_*SF = \pi_*^S$ is the classical J-homomorphism, where π_*^S denotes the stable stems (i.e., stable homotopy groups of spheres).

E-oriented GV and GV'-bundles (ξ, μ) and (ξ', μ') over X are said to be stably equivalent if $(\xi \oplus \varepsilon W, \mu \oplus \mu_0 W)$ is equivalent to $(\xi' \oplus \varepsilon W', \mu' \oplus \mu_0 W')$ for some $W \perp V$ and $W' \perp V'$ such that $V + W = V' + W'$. Write $\{\xi, \mu\}$ for the stable equivalence class of (ξ,μ) and call $\{\xi,\mu\}$ an E-oriented stable G-bundle. Let $\widetilde{K}(G;E)(X)$ denote the set of E-oriented stable G-bundles over X. $\widetilde{K}(G;E)(X)$ is classified by $B(G;E)$, the image of $[X^+, B(GV;E)]$ in $[X^+, B(G;E)]$ depends only on the dimension of V, and the product on $B(G;E)$ induces the external and internal operations $\wedge$ and $\oplus$ on E-oriented stable G-bundles given by the fibrewise smash product and Whitney sum.

A Grothendieck group $K(G;E)(X)$ can be defined, but is uninteresting. Let $Q(G;E)(X)$ denote the image of $\widetilde{K}(G;E)(X)$ in $\widetilde{K}G(X)$ under neglect of orientation. Thus

$$Q(G,E)(X) = q_*[X^+, B(G;E)] \subset [X^+, BG], \quad q: B(G;E) \to BG.$$

<u>Definition 1.1.</u> An E-orientation of G is an H-map $g: BG \to B(G;E)$ such that qg is homotopic to the identity map.

Given g, its composite with $Bj: B(G;E) \to B(F;E)$ will again be denoted by g and the following diagram will be homotopy commutative:

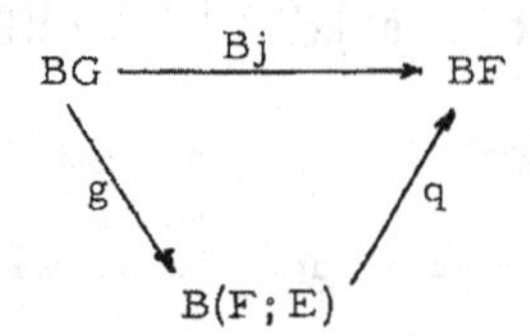

In particular, $JG(X) \subset Q(F;E)(X)$ if G admits an E-orientation.

The proof that $B(GV;E)$ classifies E-oriented GV-bundles gives an explicit universal E-oriented GV-bundle (π,θ); in an evident sense, the (π,θ) are compatible as V varies. Let $\{\xi\} \in \widetilde{K}GX$ be classified by $\alpha: X \to BG$. If $g\alpha$ factors (up to homotopy) through $B(GV;E)$, as necessarily holds for some V by the finite dimensionality of X, then the element $\{(g\alpha)^*(\pi), (g\alpha)^*(\theta)\}$ of $\widetilde{K}(G;E)(X)$ is independent of the choice of V and projects to $\{\xi\}$ in $\widetilde{K}G(X)$. If we write $\mu(g)$ for all orientations so determined by g, then the requirement that g be an H-map ensures the validity of the product formula

$$\{\xi, \mu(g)\} \wedge \{\psi, \mu(g)\} \equiv \{\xi \wedge \psi, \mu(g) \wedge \mu(g)\} = \{\xi \wedge \psi, \mu(g)\}$$

in $\widetilde{K}(G;E)(X \times Y)$ for all $\{\xi\} \in \widetilde{K}G(X)$ and $\{\psi\} \in \widetilde{K}G(Y)$.

We are particularly interested in the fine structure preserved by infinite loop maps, by which we understand maps in $H\mathcal{J}$ which are equivalent to the zeroth maps of morphisms in $H\mathcal{S}$ (see II, §2). The reader who does not share our interest may skip to the next section. The discussion there will proceed on two levels, one based on E-orientations and the other based on the following notion.

<u>Definition 1.2.</u> Let E be an E_∞ ring spectrum. An E-orientation $g: BG \to B(G;E)$ is said to be perfect if g is an infinite loop map and if $qg = 1$ as infinite loop maps.

As will be discussed in §7, there now exist homotopical proofs in several important cases of the somewhat weaker assertion that the composite of g and the natural map $B(G;E) \to B(F;E)$ is an infinite loop map whose composite with $q: B(F;E) \to BF$ is equal as an infinite loop map to $Bj: BG \to BF$. An easy Barratt-Puppe sequence argument in $\mathcal{S}$ (justified by [48, I]) shows that G admits a perfect E-orientation if and only if $e: G \to FE$ is the trivial infinite loop map.

When an E-orientation g of G is the limit over $V \subset R^\infty$ of maps $gV: BGV \to B(GV;E)$, the gV induce maps $TGV \to T(GV;E) \xrightarrow{\varepsilon} EV$. In practice, these maps define a morphism of prespectra $TG \to \nu E$ and therefore induce a morphism $\gamma: MG \to E$ of spectra. On the coordinatized level, appropriate compatibility of the gR^n, which destabilizes the statement that g is an H-map, will ensure that γ is a morphism of ring spectra in Whitehead's sense and therefore, as explained in and after II.3.4, that γ is a morphism of (weak) ring spectra in $H\mathcal{S}$. The verification that γ is actually a morphism of E_∞ ring spectra lies very much deeper and has not yet been carried out in the interesting cases.

Conversely, let $\gamma: MG \to E$ be a morphism of ring spectra and consider the following homotopy commutative diagram:

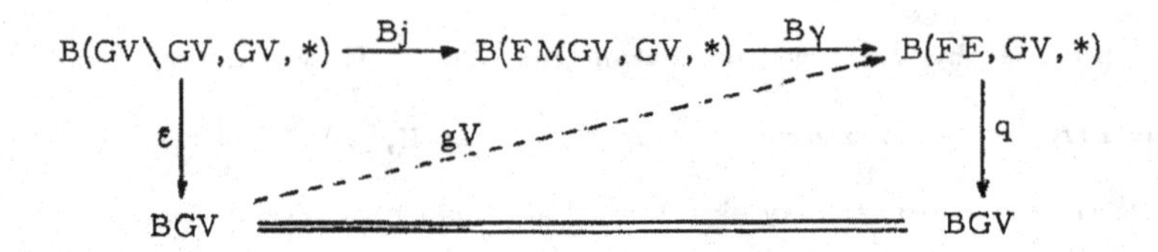

Here $j: GV\backslash GV \to FMGV$ is defined in IV.2.6. ε is a homotopy equivalence (if $GV \in \mathcal{W}$), and we define $gV = B\gamma \circ Bj \circ \varepsilon^{-1}$ for some chosen homotopy inverse ε^{-1}. Comparison of IV.2.6 with the proof of [47, 11.1] shows that gV classifies the E-oriented GV-bundle (π, μ), where $\pi: DGV \to BGV$ is the universal GV-bundle and μ is its E-orientation induced via γ from the MG-orientation $\iota: TGV \to MGV$. The gV define an E-orientation g of G by passage to limits over V, and IV.3.6 (with $\mathcal{G}$-structure ignored) shows how to recover γ from the diagram of Thom spectra over the displayed diagram of classifying spaces. By IV.2.6, $e: G \to FE$ factors through $G\backslash G$. Since $G\backslash G$ is contractible through maps of $\mathcal{L}$-spaces, by [45, 9.9 and 12.2], e is the trivial infinite loop map and g is a perfect E-orientation of G when γ is a morphism of $\mathcal{G}$-spectra.

2. Cannibalistic classes and the comparison diagram

We construct some key commutative diagrams and record a few (presumably well-known) technical lemmas here.

We write ϕ for the product on all (multiplicative) H-spaces in sight, and all H- spaces in sight will have a homotopy inverse map χ. Of course, ϕ and χ induce addition and the additive inverse on homotopy groups.

Since the stable category is additive, ϕ and χ on infinite loop spaces are infinite loop maps. For an H-space Y and two maps $f, g: X \to Y$, define $f/g = \phi(f \times \chi g)\Delta$. (We shall write $f - g$ instead of f/g when we choose to think of ϕ as additive.) f/g is an H-map if f and g are H-maps and Y is homotopy commutative (as will always be the case). If X and Y are infinite loop spaces and f and g are infinite loop maps, then f/g is an infinite loop map; in particular, $1/1 = \phi(1 \times \chi)\Delta$ is the trivial infinite loop map.

We agree to abbreviate weak homotopy equivalence to equivalence. By an equivalence of infinite loop spaces, we understand the zeroth map of a weak homotopy equivalence of spectra. The following result is an immediate consequence of the quasifibration sequence

$$G \xrightarrow{e} FE \xrightarrow{\tau} B(G;E) \xrightarrow{q} BG .$$

<u>Lemma 2.1.</u> Let g be an E-orientation of G. Then the composite

$$FE \times BG \xrightarrow{\tau \times g} B(G;E) \times B(G;E) \xrightarrow{\phi} B(G;E)$$

is an equivalence of H-spaces. If g is perfect, then $\phi(\tau \times g)$ is an equivalence of infinite loop spaces.

Our main concern in this section is with the comparison of different E-orientations of the same underlying stable G-bundle. Thus let $\{\xi, \mu\}$ and $\{\xi, \nu\}$ be E-oriented stable G-bundles over X classified by $\overline{\alpha}$ and $\overline{\beta}$. Since $q\overline{\alpha} \simeq q\overline{\beta} : X \to BG$ and since $q: B(G;E) \to BG$ commutes up to homotopy with ϕ and χ,

$$q(\overline{\alpha}/\overline{\beta}) = q\phi(\overline{\alpha} \times \chi\overline{\beta})\Delta \simeq \phi(q\overline{\alpha} \times \chi q\overline{\beta})\Delta \simeq \phi(1 \times \chi)\Delta q\overline{\alpha} \simeq *.$$

Since $\tau: FE \to B(G;E)$ is canonically equivalent to the fibre of q, this null homotopy determines a map $\delta: X \to FE$ such that $\tau\delta \simeq \overline{\alpha}/\overline{\beta}$. Clearly $\phi(\tau\delta \times \overline{\beta})\Delta \simeq \overline{\alpha}$. Since $\tau\delta$ classifies $\{\varepsilon_0, \delta\}$, where the unit $\delta \in E^0(X^+)$ is regarded as an orientation of the trivial $G(0)$-bundle $\varepsilon_0: X \times S^0 \to X$ (with Thom complex X^+), and since, by the explicit definition of $\delta \oplus \nu$ in III.1.5,

$$\{\varepsilon_0, \delta\} \oplus \{\xi, \nu\} = \{\varepsilon_0 \oplus \xi, \delta \oplus \nu\} = \{\xi, \delta \cup \nu\},$$

we conclude that $\{\xi, \mu\} = \{\xi, \delta \cup \nu\}$.

In our applications of this difference construction, we shall be given a classifying space Y for some class of bundles with additional structure and we shall be given maps $a, b: Y \to B(G; E)$ such that $qa \simeq qb$. The classifying maps $\bar{\alpha}$ and $\bar{\beta}$ above will be $a\gamma$ and $b\gamma$ for a classifying map $\gamma: X \to Y$ and δ will be $d\gamma$ for a map $d: Y \to FE$ such that $\tau d \simeq a/b$. Note that the null homotopy $q(a/b) \simeq *$, hence also d and the homotopy $\tau d \simeq a/b$, are explicitly and canonically determined by the homotopies

$$qa \simeq qb, \quad q\phi \simeq \phi(q \times q), \quad q\chi \simeq \chi q, \quad \text{and} \quad \phi(1 \times \chi)\Delta \simeq *.$$

If a and b are H-maps, then so is d. If E is an E_∞ ring spectrum, a and b are infinite loop maps, and $qa = qb$ as infinite loop maps, then d is an infinite loop map and $\tau d = a/b$ as infinite loop maps.

The theory of cannibalistic classes fits nicely into this framework. Let $\psi: E \to E$ be a map of ring spectra. Then $qB\psi = q$, $B\psi: B(G; E) \to B(G; E)$, and there results a canonical H-map $c(\psi): B(G; E) \to FE$ such that $\tau c(\psi) \simeq B\psi/1$; if $B\psi$ is an infinite loop map, then so is $c(\psi)$. We call $c(\psi)$ the universal cannibalistic class determined by ψ. If $\{\xi, \mu\}$ is an E-oriented stable G-bundle over X classified by $\bar{\alpha}$, write

$$c(\psi) = c(\psi)\{\xi, \mu\} = \bar{\alpha}^* c(\psi) \in E^0(X^+).$$

Define ψ on $\widetilde{K}(G; E)(X)$ by $\psi\{\xi, \mu\} = \{\xi, \psi\mu\}$ and note that $\psi\{\xi, \mu\}$ is classified by $B\psi \circ \bar{\alpha}$. The discussion above shows that

$$\psi\{\xi, \mu\} = \{\xi, c(\psi) \cup \mu\} \in \widetilde{K}(G; E)(X).$$

Of course, given an E-orientation g of G, we can define cannibalistic classes $\rho = \rho\{\xi\}$ for stable G-bundles $\{\xi\}$ by $\rho\{\xi\} = c(\psi)\{\xi, \mu(g)\}$. The fact that these classes are represented by the composite $c(\psi)g$ on the universal level will play an essential role in our theory. We note next that $\psi/1$ also factors through $c(\psi)$.

Proposition 2.2. The following diagram is homotopy commutative:

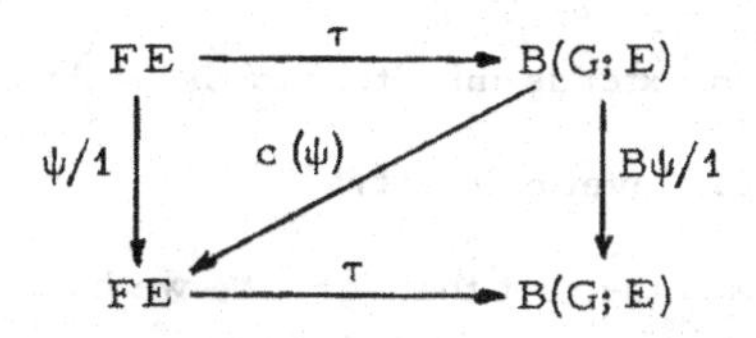

If E is an E_∞ ring spectrum and ψ on FE and $B\psi$ on $B(G;E)$ are infinite loop maps, then the diagram determines a corresponding diagram on the level of spectra.

Proof. $B\psi/1 \simeq \tau \circ c(\psi)$ by the definition of $c(\psi)$, and $(B\psi/1)\tau \simeq \tau(\psi/1)$ since $B\psi \circ \tau = \tau \circ \psi$. By construction, $c(\psi)$ is natural in G. When $G = e$ is trivial, $\tau = 1$ and $\psi = B\psi$ on $FE = B(e;E)$, hence $\psi/1 \simeq c(\psi)$. Now $c(\psi) \circ \tau \simeq \psi/1$ follows by an obvious diagram chase. The last statement holds by the general observations above.

The main interest often lies not in an E-oriented bundle theory but in its relationship to a larger bundle theory. Thus assume given morphisms $H \xrightarrow{i} G \xrightarrow{j} F$ of monoid-valued $\mathcal{J}_*$-functors and assume given an E-orientation g of H. We also write g for its composite with $Bi: B(H;E) \rightarrow B(G;E)$. There are now two natural E-orientations of G-trivialized stable H-bundles in sight, namely those given by the maps

$$G/H \xrightarrow{q} BH \xrightarrow{g} B(H;E) \quad \text{and} \quad G/H \xrightarrow{Be} B(H;E).$$

Their quotient H-map gq/Be factors as τf, where $f: G/H \rightarrow FE$ is an H-map by virtue of Lemma 2.1 and is an infinite loop map if g is perfect. With these notations, we have the following result.

Proposition 2.3. The first three squares of the following "comparison diagram" are homotopy commutative:

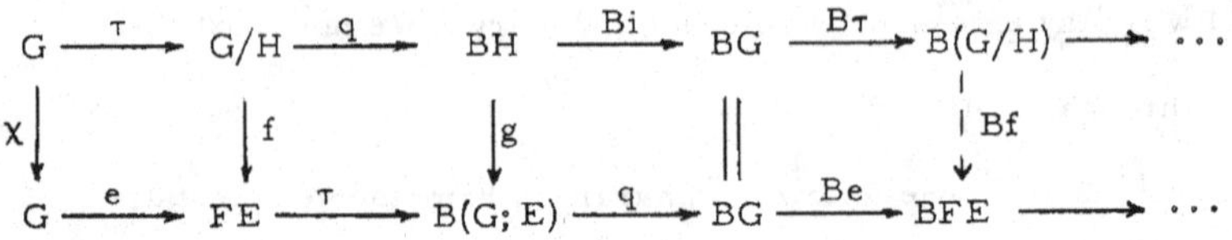

$$\begin{array}{ccccccccccc}
G & \xrightarrow{\tau} & G/H & \xrightarrow{q} & BH & \xrightarrow{Bi} & BG & \xrightarrow{B\tau} & B(G/H) & \longrightarrow & \cdots \\
\downarrow{\scriptstyle \chi} & & \downarrow{\scriptstyle f} & & \downarrow{\scriptstyle g} & & \| & & \downarrow{\scriptstyle Bf} & & \\
G & \xrightarrow{e} & FE & \xrightarrow{\tau} & B(G;E) & \xrightarrow{q} & BG & \xrightarrow{Be} & BFE & \longrightarrow & \cdots
\end{array}$$

If g is perfect, the diagram extends infinitely to the right and determines a corresponding diagram on the level of spectra.

Proof. By III. 2.6 and the fact that $q\tau = *$, we have

$$\tau f \tau \simeq \phi(gq \times \chi Be)\Delta\tau \simeq \chi(Be)\tau \simeq \chi\tau e \simeq \tau\chi e \simeq \tau e \chi ,$$

where the first map τ takes FE to $B(H;E)$ and is thus the injection of a factor by Lemma 2.1. Therefore $f\tau \simeq e\chi$. We have the commutative diagram

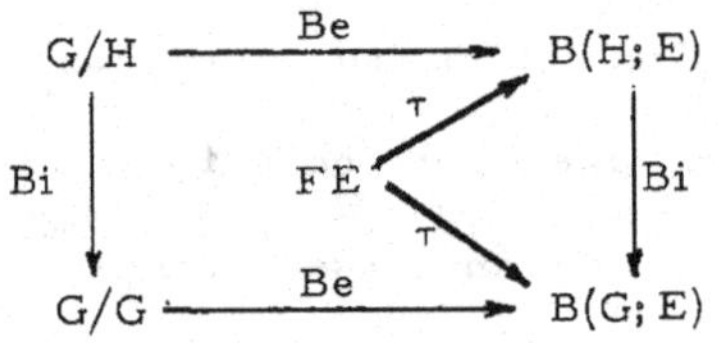

$$\begin{array}{ccccc}
G/H & & \xrightarrow{Be} & & B(H;E) \\
 & & & \nearrow{\scriptstyle \tau} & \\
\downarrow{\scriptstyle Bi} & & FE & & \downarrow{\scriptstyle Bi} \\
 & & & \searrow{\scriptstyle \tau} & \\
G/G & & \xrightarrow{Be} & & B(G;E)
\end{array}$$

in which G/G is contractible through infinite loop maps. Thus $BiBe$ is the trivial infinite loop map and (since Bi commutes with ϕ and χ) we have

$$\tau f = (Bi)\tau f = (Bi)\,\phi(gq \times \chi Be)\Delta \simeq (Bi)gq = gq .$$

The third square homotopy commutes by the definition of an orientation.

The reader familiar with Barratt-Puppe sequences will wonder why the sign given by χ appears. If one writes down explicit equivalences of the two rows with honest fibration sequences, starting from BG and working left, one produces two homotopy equivalences $G \to \Omega BG$. These turn out to differ by χ. Of course, given g, Barratt-Puppe sequence arguments (e.g. [48, I§2]) produce a map f, not uniquely determined, such that the two left squares homotopy commute. Conversely, given Bf such that the right square homotopy commutes, there exists g such that the rest of the diagram homotopy commutes. These statements remain valid with H and G replaced by HV and GV, V finite dimensional, in which case the explicit construction of f fails for lack of an H-space structure on $B(GV; E)$.

We shall need some observations concerning localizations of the comparison diagram at a set of primes T. We restrict to the integrally oriented case in order to deal with connected spaces.

The localization E_T is again a commutative ring spectrum, and $SF(E_T) \simeq (SFE)_T$. I do not know if localizations of E_∞ ring spectra are E_∞ ring spectra, but any infinite loop space information derived from the E_∞ structure is preserved under localization. We write λ generically for localization at T.

Lemma 2.4. For any G, the following composite is a localization at T:

$$B(SG; E) \xrightarrow{B\lambda} B(SG; E_T) \xrightarrow{\lambda} B(SG; E_T)_T .$$

When $G = F$, $BSF \simeq (BSF)_T \times BSF[T^{-1}]$ and the map

$$(\lambda, q)\colon B(SF; E_T) \rightarrow B(SF; E)_T \times BSF[T^{-1}]$$

is an equivalence (of infinite loop spaces if E is an E_∞ ring spectrum).

Proof. In view of the following homotopy commutative diagram, this is immediate from the fact that localization preserves fibrations of connected H-spaces:

$$\begin{array}{ccccccccc}
SG & \xrightarrow{e} & SFE & \xrightarrow{\tau} & B(SG; E) & \xrightarrow{q} & BSG & \xrightarrow{Be} & BSFE \\
\Vert & & \downarrow{\scriptstyle\lambda} & & \downarrow{\scriptstyle B\lambda} & & \Vert & & \downarrow{\scriptstyle\lambda} \\
SG & \xrightarrow{e} & SFE_T & \xrightarrow{\tau} & B(SG; E_T) & \xrightarrow{q} & BSG & \xrightarrow{Be} & BSFE_T \\
\downarrow{\scriptstyle\lambda} & & \Vert & & \downarrow{\scriptstyle\lambda} & & \downarrow{\scriptstyle\lambda} & & \Vert \\
SG_T & \xrightarrow{e_T} & SFE_T & \xrightarrow{\tau_T} & B(SG; E_T)_T & \xrightarrow{q_T} & BSG_T & \xrightarrow{(Be)_T} & BSFE_T
\end{array}$$

When $G = F$ in the comparison diagram (in which case we rename $H = G$), f can sometimes be intrinsically characterized in terms of g.

Lemma 2.5. Let g be an E_T-orientation of G and assume that the following two conditions hold.

(i) $H_*(SF/SG)$, $H_*(SFE)$, and $\pi_*(SFE)$ have no T-torsion.

(ii) $H_*(SF/SG)$ and $H_*(SFE; Q)$ are of finite type (over Z and Q, respectively).

Then $f: SF/SG \to SFE_T$ is the unique H-map such that the second square of the comparison diagram homotopy commutes.

Proof. Given another such H-map f', f/f' factors through SF and therefore induces the zero map on homotopy. Thus $f_* = f'_*$ on homotopy. As pointed out to me by Frank Adams, the following pair of lemmas complete the proof.

Lemma 2.6. Let X and Y be connected homotopy associative H-spaces, with $H_*(X; Q)$ of finite type. If two H-maps $X \to Y$ induce the same homomorphism $\pi_*X \otimes Q \to \pi_*Y \otimes Q$, then they induce the same homomorphism $H_*(X; Q) \to H_*(Y; Q)$.

Proof. By Milnor and Moore [50, Appendix], the Hurewicz homomorphism $h: \pi_*X \to H_*X$ induces a monomorphism upon tensoring with Q, and the image of this monomorphism generates $H_*(X; Q)$ as an algebra.

Lemma 2.7. Let X be a connnected CW-complex and let Y be a connected homotopy associative T-local H-space. Assume that the following two conditions hold.

(i) H_*X, H_*Y, and π_*Y have no T-torsion.

(ii) H_*X and $H_*(Y; Q)$ are of finite type (over Z and Q, respectively).

Then two maps $f, f': X \to Y$ are homotopic if they induce the same homomorphism $H_*(X; Q) \to H_*(Y; Q)$.

Proof. The hypotheses imply that f and f' induce the same homomorphism on integral homology. Suppose that f and f' are homotopic on the (n-1)-skeleton of X. If $k \in H^n(X; \pi_n Y)$ is the obstruction to the extension of the restriction to the (n-2)-skeleton of a given homotopy, then

$$h\langle k, x\rangle = f_*(x) - f'_*(x) = 0 \quad \text{for } x \in H_* X,$$

where $\langle\ ,\ \rangle : H^n(X; \pi_n Y) \otimes H_n X \to \pi_n Y$ is the Kronecker product (the first equality holding by explicit chain level calculation from the definitions). Since $h: \pi_* Y \to H_* Y$ is a monomorphism, $\langle k, x\rangle = 0$. Since $\mathrm{Ext}(H_{n-1} X, \pi_n Y) = 0$ (because $H_* X$ is of finite type, $H_* X$ and $\pi_* Y$ have no T-torsion, and $\pi_* Y$ is a Z_T-module), $k = 0$ by the universal coefficient theorem. (See VIII. 1. 1 for a simpler proof when $Y = \Omega Z$.)

The following analog of Lemma 2.7 was also pointed out to me by Frank Adams.

Lemma 2.8. Let X and Y be spaces of the homotopy type of BSO_T, where T is any set of primes. Then two maps $f, f': X \to Y$ are homotopic if they induce the same homomorphism $H_*(X; Q) \to H_*(Y; Q)$.

Proof. Let $A = \varinjlim BT(n)$, where $T(n)$ is a maximal torus in $SO(n)$, and let $i: A \to BSO$ be the evident inclusion. Consider the diagram

$$\begin{array}{ccccc} & & H^{**}(BSO; Q) & \xrightarrow{i^{**}} & H^{**}(A; Q) \\ & & \uparrow \mathrm{ch} & & \uparrow \mathrm{ch} \\ \widetilde{KO}(BSO) & \xrightarrow{c} & \widetilde{KU}(BSO) & \xrightarrow{i^*} & \widetilde{KU}(A) \end{array}$$

Clearly i^{**}, and ch on $\widetilde{KU}(A)$, are monomorphisms. By Atiyah and Segal [14] and an inverse limit argument, i^* is a monomorphism. Thus ch on $\widetilde{KU}(BSO)$ is a monomorphism. By Anderson [9, p. 38] or [14], c is also a monomorphism (in fact an isomorphism). These statements remain true after localization of BSO and the representing spaces BO and BU at T. It follows that f and f' induce the same map of localized real K-theory and therefore, since $BO \simeq BSO \times BO(1)$, that $f^* = (f')^* : [Y, Y] \to [X, Y]$.

In VIII, we shall use the fact that this result remains valid, by the same proof, for completions at T. The following observation explains why rational information determines the behavior with respect to self-maps of the 2-torsion in $\pi_* BSO_T$.

<u>Lemma 2.9.</u> Let $f: BSO_T \to BSO_T$ be a map, where $2 \in T$. Let $a_j \in Z_T$ be such that $f_*(x) = a_j x$ for all $x \in \pi_{4j} BSO_T \cong Z_T$.

(i) If $0 \neq y \in \pi_2 BSO_T$, then $f_*(y) = a_1 y$.

(ii) If $0 \neq y \in \pi_{8j+k} BSO_T$, $j \geq 1$ and $k = 1$ or $k = 2$, then $f_*(y) = a_{2j} y$.

In both statements, the coefficients are understood to be reduced mod 2.

<u>Proof.</u> For (i), let $p_n \in H_n(BSO; Z_2)$ be the unique non-zero primitive element and recall that

$$Sq_*^1 p_3 = p_2 \text{ , } Sq_*^2 p_5 = p_3 \text{ , and } Sq_*^1 p_5 = p_4 \text{ .}$$

Clearly $f_*(y) = 0$ if and only if $f_*(p_2) = 0$, and the displayed equations show that $f_*(p_2) = 0$ if and only if $f_*(p_4) = 0$. By an obvious argument with the cover $BSpin_T \to BSpin_T$ of α, $f_*(p_4) = 0$ if and only if $a_1 \equiv 0 \bmod 2$. For (ii), simply recall that if $x: S^{8j} \to BSO_T$ generates $\pi_{8j} BSO_T$, then $y = x \circ \eta$ or $y = x \circ \eta^2$, where $\eta^k : S^{8j+k} \to S^{8j}$ is the non-trivial map.

Finally, the following result, which I learned from Anderson and Snaith, implies that Lemma 2.8 remains true when $X \simeq BSpin_T$ and $Y \simeq BSO_T$.

<u>Lemma 2.10.</u> The natural map $\pi: BSpin \to BSO$ induces isomorphisms on real and complex K-theory. Therefore $\pi^* : [BSO, BSO] \to [BSpin, BSO]$ is an isomorphism.

Proof. There are no phantom maps here (since $KU^{-1}(BG) = 0$ and $KO^{-1}(BG)$ is finite for a compact Lie group G), and it suffices to consider the inverse systems of completed (real and complex) representation rings of Spin(n) and SO(n). We may as well consider only odd n, where

$$RO(SO(2m+1)) = R(SO(2m+1)) = P\{\lambda_1, \ldots, \lambda_m\}$$

maps surjectively to $R(SO(2m-1))$ and injectively to

$$RO(Spin(2m+1)) \subset R(Spin(2m+1)) = P\{\lambda_1, \ldots, \lambda_m, \Delta_m\}.$$

Since Δ_m restricts to $2\Delta_{m-1}$ in $R(Spin(2m-1))$ by a check of characters, the exceptional spinor representations make no contribution to the inverse limit.

3. The kO-orientation of Spin and the J-theory diagram

The main examples involve K-theory. I do not know if there exist E_∞ ring spectra which represent KO^* and KU^*, but explicit E_∞ ring spectra kO and kU which represent the associated connective theories are constructed in VII. Write $BU_\otimes = SFkU$ and $BO_\otimes = SFkO$; informally, these infinite loop spaces are the 1-components of $BU \times Z$ and $BO \times Z$.

Lemma 3.1. For G = O and G = U, $BG_\otimes$ is equivalent as an infinite loop space to $BG(1) \times BSG_\otimes$.

Proof. $BO(1) = K(Z_2, 1)$ and $BU(1) = K(Z, 2)$ admit unique infinite loop space structures, the natural map $BG_\otimes \to BG(1)$ obtained by killing π_1 or π_2 (i.e., w_1 or c_1) is automatically an infinite loop map, and we define $BSG_\otimes$ and $\pi: BSG_\otimes \to BG_\otimes$, as an infinite loop space and map, to be its fibre. By VI.4.5, the natural inclusion $\eta: BG(1) \to BG_\otimes$ (which classifies the canonical line bundle) is an infinite loop map. The composite

$$BG(1) \times BSG_{\otimes} \xrightarrow{\eta \times \pi} BG_{\otimes} \times BG_{\otimes} \xrightarrow{\phi} BG_{\otimes}$$

is the desired equivalence.

Atiyah, Bott, and Shapiro [13] have constructed a kO-orientation of Spin and a kU-orientation of $Spin^c$. Thus, if Spin maps to G, we have well-defined H-maps

$$g: BSpin \to B(G; kO) \quad \text{and} \quad f: G/Spin \to BO_{\otimes}.$$

The fibre of the quasi-fibration $B(SO, Spin, *) \to B(SO, SO, *)$ is precisely $BZ_2 = BO(1)$, and this gives an explicit equivalence $i: BO(1) \to SO/Spin$. Similarly, if $Spin^c$ maps to G, we have H-maps

$$g: BSpin^c \to B(G; kO) \quad \text{and} \quad f: G/Spin^c \to BO_{\otimes},$$

and we have an explicit equivalence $i: BU(1) \to SO/Spin^c$.

Proposition 3.2. The following composites are equivalences:

$$BSO_{\otimes} \times BSpin \xrightarrow{\tau \times g} B(SO; kO) \times B(SO; kO) \xrightarrow{\phi} B(SO; kO)$$

and

$$BSU_{\otimes} \times BSpin^c \xrightarrow{\tau \times g} B(SO; kU) \times B(SO; kU) \xrightarrow{\phi} B(SO; kU).$$

Proof. The result is proven by easy chases of the relevant comparison diagram of Proposition 2.3. In the real case, the salient facts are that BSpin is equivalent to the fibre of $w_2: BSO \to K(Z_2, 2)$ and that $e_*: \pi_1 SO \to \pi_1 BO_{\otimes}$ is an isomorphism (because the obstruction $w_2(\xi; kO)$ of IV, § 3 can be non-zero or by direct calculation on mod 2 homology). In the complex case, the salient facts are that $BSpin^c$ is equivalent to the fibre of $w_3: BSO \to K(Z, 3)$ or (equivalently by a comparison of fibrations) to the fibre of $w_2 \otimes 1 + 1 \otimes \iota : BSO \times K(Z,2) \to K(Z_2, 2)$ and that $\pi_2 BSpin^c = Z$.

Corollary 3.3. The composite $fi: BO(1) \to SO/Spin \to BO_{\otimes}$ is homotopic to the natural inclusion $\eta: BO(1) \to BO_{\otimes}$.

Proof. Consider the following diagram, where $\zeta: BO(1) \to SO$ is any map which is non-trivial on π_1:

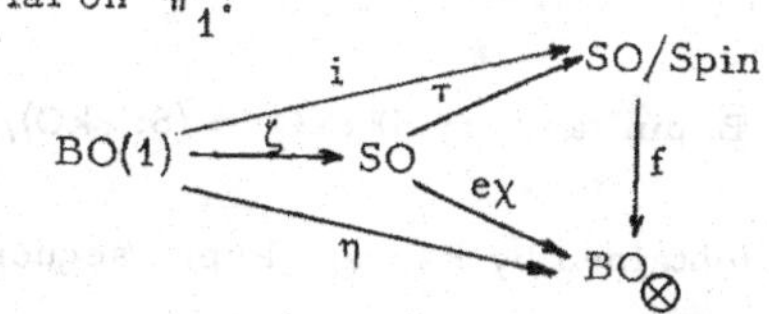

$\tau\zeta \simeq i$ and $e\chi\zeta \simeq \eta$ because both composites induce isomorphisms on π_1 and because the component of e in $BSO_{\otimes}$ is null homotopic by the splitting of $B(SO;kO)$. $f\tau \simeq e\chi$ by the comparison diagram.

The argument fails in the (less interesting) complex case, and I have not verified whether or not $fi \simeq \eta$ in that case.

We turn to the study of kO-oriented spherical fibrations. The rest of the section will be concerned with the construction and analysis of the "J-theory diagram", which is obtained by superimposing diagrams involving cannibalistic classes and the Adams conjecture on an elaboration of the comparison diagram for

$$g\colon BSpin \to B(SF;kO) \quad \text{and} \quad f\colon SF/Spin \to BO_{\otimes}.$$

Of course, this map f restricts on $SO/Spin$ to that just discussed, and we have the following observation.

<u>Lemma 3.4.</u> The natural map $SF/Spin \to F/O$ and the composite of f and $w_1\colon BO_{\otimes} \to BO(1)$ are the components of an equivalence $SF/Spin \to F/O \times BO(1)$ of infinite loop spaces.

Define $O_{\otimes} = \Omega BO_{\otimes}$ and $(SF;kO) = \Omega B(SF;kO)$. By abuse, write G and ΩBG interchangeably when $G = Spin, SF$, etc. For a map $\theta\colon X \to Y$, write π and ι generically for the projection $F\theta \to X$ from the (homotopy theoretic) fibre and for the inclusion $\Omega Y \to F\theta$. Define $(SF;kO)/Spin$ to be the fibre of f (which is equivalent to the fibre of g). This space classifies stable Spin-bundles with trivializations as kO-oriented stable spherical fibrations. Just as if $\Omega g\colon Spin \to (SF;kO)$ were derived from a morphism of monoid-valued $\mathcal{J}_*$-functors, write

$$q: (SF; kO)/Spin \to BSpin \text{ and } \tau: (SF; kO) \to (SF; kO)/Spin$$

for $q\pi$ and for a map τ (obtained by Barratt-Puppe sequence arguments) such that $\pi\tau \simeq \tau\Omega q$ and $\tau\Omega\tau \simeq \iota$ in the diagram on the following page. With these notations, the solid arrow portion of this diagram exists and is homotopy commutative by Proposition 2.3.

At the right of the diagram, $BSpin_{\otimes}$ is defined (as an infinite loop space) to be the fibre of $w_2: BSO_{\otimes} \to K(Z_2, 2)$.

We claim that, with dotted arrows inserted, this J-theory diagram exists and is homotopy commutative when $r \geq 2$ and all spaces in sight are localized away from r. To see this, first recall the following calculations of Adams [1, 5.1 and 5.2]. (See also Lemma 2.9.)

<u>Theorem 3.5.</u> ψ^r is a natural ring homomorphism on $KO(X)$ such that $\psi^r \xi = \xi^r$ on line bundles ξ. Let $x \in \widetilde{KO}(S^i) = \pi_i BO$, $i > 0$. If $i = 4j$, $\psi^r x = r^{2j}x$; if $i \equiv 1$ or $2 \bmod 8$ and r is odd, $\psi^r x = x$.

It follows by II.3.15 that ψ^r determines a morphism of ring spectra $kO[1/r] \to kO[1/r]$ and thus that $c(\psi^r): B(SF; kO) \to BO_{\otimes}$ is defined away from r. Since $e_*: \pi_i SF \to \pi_i BO_{\otimes}$ is an isomorphism for $i = 1$ (by Corollary 3.3) and for $i = 2$ (by translation from $Q_0S^0 \to BO$, where the π_2's are generated by the smash and tensor product squares of the generators of the π_1's), $B(SF; kO)$ is 2-connected. Therefore $c(\psi^r)$ lifts uniquely to a map, still denoted $c(\psi^r)$, into $BSpin_{\otimes}$. BC^r is defined to be the fibre of this map, and C^r is defined to be ΩBC^r. For r even, $BO \simeq BSO \simeq BSpin$ localized away from r. For r odd, Theorem 3.5 implies that $\psi^r/1: BO_{\otimes} \to BO_{\otimes}$ is trivial on $BO(1)$ (because the square of a line bundle is trivial) and annihilates $\pi_2 BSO_{\otimes}$. The splitting $SO \simeq Spin \times RP^{\infty}$ determined by the fibration $Spin \to SO \to SO/Spin$ and any map $\zeta: RP^{\infty} \to SO$ which is non-trivial on π_1

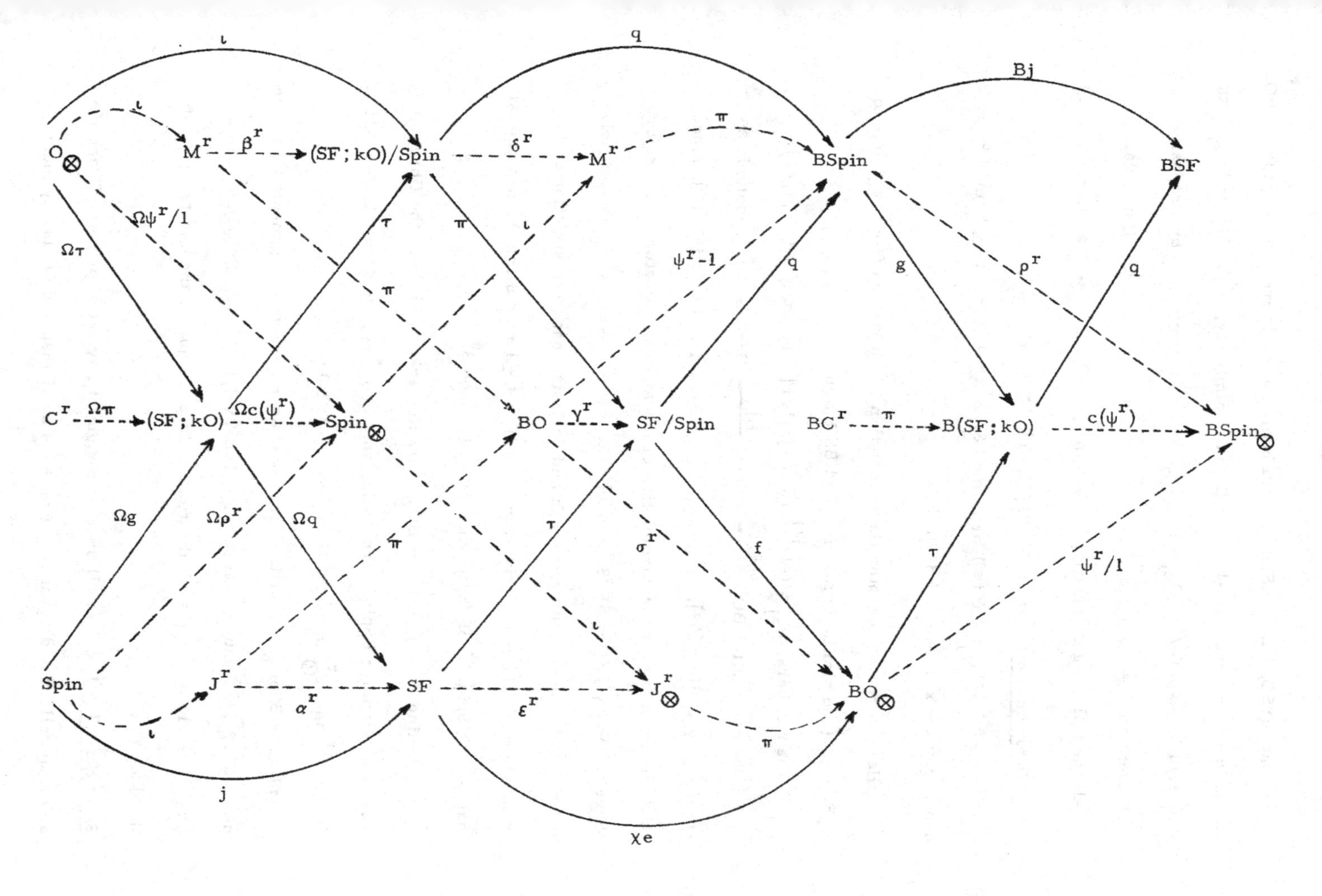
O⊗
M^r
β^r
(SF; kO)/Spin
δ^r
M^r
BSpin
BSF
ι
q
Bj
π
Ωψ^r/1
Ωτ
τ
ψ^r-1
g
ρ^r
C^r
Ωπ
(SF; kO)
Ωc(ψ^r)
Spin⊗
BO
γ^r
SF/Spin
BC^r
B(SF; kO)
c(ψ^r)
BSpin⊗
Ωg
Ωρ^r
Ωq
σ^r
f
ψ^r/1
Spin
J^r
α^r
SF
ε^r
J^r⊗
BO⊗
j
χe

shows that $q: SO/Spin \to BSpin$ is null homotopic. Therefore $\psi^r - 1: BO \to BO$ lifts uniquely to a map $\psi^r - 1: BO \to BSpin$. Similarly, $\psi^r/1: BO_\otimes \to BO_\otimes$ lifts uniquely to a map $\psi^r/1: BO_\otimes \to BSpin_\otimes$, and $\psi^r/1 \simeq c(\psi^r)\tau$ by Proposition 2.2.

Define $\rho^r = c(\psi^r)g$; thus ρ^r is the Adams-Bott cannibalistic class. Recall the following calculations of Adams [3, p. 166]. (See also Lemma 2.9.)

Theorem 3.6. Let $x \in \pi_i BSpin[1/r]$, $i > 0$. If $i = 4j$, $\rho^r x = 1 + \frac{1}{2}(r^{2j} - 1)\alpha_{2j} x$; if $i \equiv 1$ or 2 mod 8, $\rho^r x = 1$ for $r \equiv \pm 1$ mod 8 and $\rho^r x = 1 + x$ for $r \equiv \pm 3$ mod 8.

Here $x \to 1 + x$ denotes the translation isomorphism from $\pi_* BSpin$ to $\pi_* BSpin_\otimes$ (and similarly for BO and BSO below).

The numbers $\alpha_{2j} = (-1)^{j+1} B_j/2j \in Z[1/r]$ are analyzed in [3, § 2].

The composite $BO \xrightarrow{\psi^r - 1} BSpin \xrightarrow{Bj} BSF$ is null homotopic away from r, by Quillen [58], Sullivan [73], or Becker and Gottlieb [17], since this statement is just a reformulation of the Adams conjecture. Therefore there exists $\gamma^r: BO \to SF/Spin$ such that $q\gamma^r \simeq \psi^r - 1$. γ^r is not uniquely determined. In particular, we can and do insist that its restriction to the translate of BO(1) be the trivial map when $r \equiv \pm 1$ mod 8 and the non-trivial map to $SO/Spin \subset SF/Spin$ when $r \equiv \pm 3$ mod 8.

Define $\sigma^r = f\gamma^r: BO \to BO_\otimes$. By Remarks 3.7 below, the fibres of σ^r and of $\rho^r: BSpin \to BSpin_\otimes$ are equivalent; by abuse, we denote both by M^r. Define J^r and $J^r_\otimes$ to be the fibres of $\psi^r - 1: BO \to BSpin$ and of $\psi^r/1: BO_\otimes \to BSpin_\otimes$. Standard Barratt-Puppe sequence arguments then give maps α^r, β^r, δ^r, and ε^r such that $(1, \alpha^r, \gamma^r)$, $(1, \beta^r, \gamma^r)$, $(\Omega c(\psi^r), \delta^r, 1)$ and $(\Omega c(\psi^r), \varepsilon^r, 1)$ are maps of fibrations. This completes the construction of the diagram.

Remarks 3.7. Localize all spaces in sight at any set of primes T and consider the following diagram, where θ is an H-map and ξ is any map:

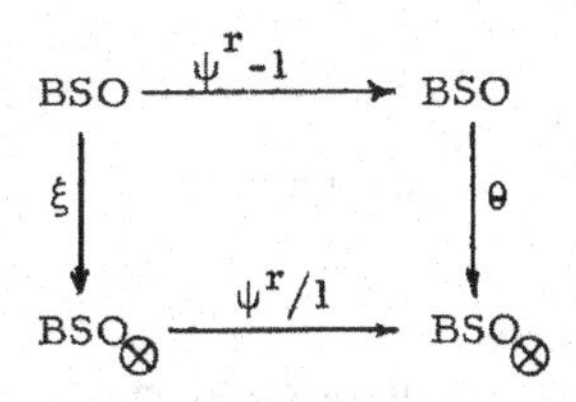

If $\xi = \theta$, the diagram is homotopy commutative by Lemma 2.8 (or, if $2 \nmid T$, Lemma 2.7) since, regardless of what θ does, $\theta(\psi^r-1)$ and $(\psi^r/1)\theta$ induce the same map on rational homology by Lemma 2.6 and Theorem 3.5. In particular, away from s, $\rho^s(\psi^r-1) \simeq (\psi^r/1)\rho^s$. Conversely, if the diagram is homotopy commutative, then $\xi_* = \theta_*$ on rational homology (by the known behavior of ψ^r-1 and $\psi^r/1$ on rational homology) and thus $\xi \simeq \theta$ by the cited lemmas. In particular, away from 2 and r, the maps σ^r and ρ^r of the J-theory diagram are homotopic. Away from r, Adams [3] has constructed an H-map $\rho^r: BSO \simeq BSO_\otimes$ which has the cannibalistic class ρ^r as 2-connective cover, and the existence of such a map also follows directly from Lemma 2.10. Clearly Adams' map and the simply connected cover of σ^r induce the samd homomorphism on rational homology and are thus homotopic.

For clarity, we shall retain the now duplicative notations σ^r and ρ^r, since what is most important about σ^r is not its homotopy class but its location in the J-theory diagram. That diagram and the remarks above give the following result, which should be compared with Proposition 2.2.

<u>Proposition 3.8</u>. Away from r, the following diagram is homotopy commutative:

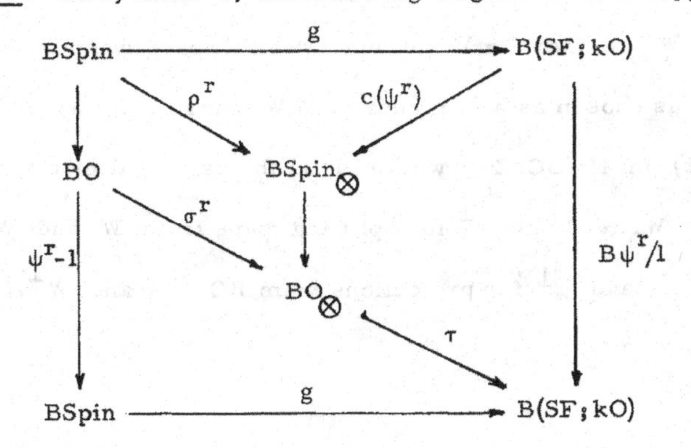

Corollary 3.9. Let $x \in \pi_i BO[1/r]$, $i > 0$. If $i = 4j$, $\sigma^r x = 1 + \frac{1}{2}(r^{2j} - 1)\alpha_{2j} x$; if $i \equiv 1$ or $2 \bmod 8$, $\sigma^r x = 1$ for $r \equiv \pm 1 \bmod 8$ and $\sigma^r x = 1 + x$ for $r \equiv \pm 3 \bmod 8$.

Proof. For $i \geq 4$, this is immediate from Theorem 3.6. For $i = 2$ it follows from Lemma 2.9. For $i = 1$, it holds by our choice of γ^r and Corollary 3.3.

4. Local analysis of the J-theory diagram

We shall analyze the localization of the J-theory diagram at each prime p, with r chosen so as to yield maximum information. Let $r(2) = 3$. For p odd, let $r(p)$ be any chosen prime power the image of which in the ring Z_{p^2} generates its group of units. This choice of $r(p)$ is motivated by the following facts [3, § 2]. Let $Z_{(p)}$ denote the localization of the integers at p.

Lemma 4.1. Let $r = r(p)$. In $Z_{(p)}$, $\frac{1}{2}(r^{2j}-1)\alpha_{2j}$ is a unit if $p = 2$ or if $p > 2$ and $2j \equiv 0 \bmod (p-1)$, while $r^{2j}-1$ is a unit if $p > 2$ and $2j \not\equiv 0 \bmod (p-1)$.

Throughout this section, unless otherwise specified, all spaces in sight are assumed to be localized at p and r denotes $r(p)$. We write BO, BSO, and BSpin interchangeably when p is odd. Recall from Adams [6, Lecture 4] or Peterson [53] that, at odd p, BO splits as an infinite loop space as $W \times W^{\perp}$, where $\pi_i W = 0$ unless $i = 2j(p-1)$ when $\pi_i W = Z_{(p)}$. The letter W is chosen as a reminder that W carries the Wu classes $w_j = \Phi^{-1}P^j\Phi(1)$ in $H^*(BO; Z_p)$, where Φ is the canonical mod p Thom isomorphism. Write ν and $\nu^{\perp}$ for splitting maps from W and $W^{\perp}$ to BO and write ω and $\omega^{\perp}$ for projections from BO to W and $W^{\perp}$.

Adams and Priddy [8], using Adams spectral sequence techniques, have recently proven the following characterization of BSO as an infinite loop space.

Theorem 4.2. There exists one and, up to isomorphism in the stable category, only one connective spectrum the zeroth space of which is equivalent to the localization (or completion) of BSO (or of BSU) at any given prime.

For p odd, we again write $\nu, \nu^{\perp}, \omega$, and $\omega^{\perp}$ for infinite loop maps which split X as $W \times W^{\perp}$ for any (p-local) infinite loop space X equivalent to BSO. (Examples include $BO_{\otimes}$ and F/Top.) The requisite splitting exists by the theorem and the splitting of BSO. We shall need the following immediate consequence of Lemmas 2.6 and 2.7.

Lemma 4.3. Let $p > 2$ and let $\theta: X \to Y$ be an H-map between H-spaces of the same homotopy type as BO. Then

$$\theta\nu \simeq \nu\omega\theta\nu : W \to Y \quad \text{and} \quad \theta\nu^{\perp} \simeq \nu^{\perp}\omega^{\perp}\theta\nu^{\perp} : W^{\perp} \to Y.$$

The following basic result summarizes information contained in Theorems 3.5 and 3.6, Corollary 3.9, and Lemma 4.1.

Theorem 4.4. At $p = 2$, $\sigma^3: BO \to BO_{\otimes}$ and $\rho^3: BSpin \to BSpin_{\otimes}$ are equivalences. At $p > 2$, the following composites are equivalences:

$$W \xrightarrow{\nu} BO \xrightarrow{\sigma^r} BO_{\otimes} \xrightarrow{\omega} W, \quad W^{\perp} \xrightarrow{\nu^{\perp}} BO \xrightarrow{\psi^r - 1} BO \xrightarrow{\omega^{\perp}} W^{\perp}$$

$$W \xrightarrow{\nu} BO \xrightarrow{\rho^r} BO_{\otimes} \xrightarrow{\omega} W, \quad W^{\perp} \xrightarrow{\nu^{\perp}} BO_{\otimes} \xrightarrow{\psi^r/1} BO_{\otimes} \xrightarrow{\omega^{\perp}} W^{\perp}$$

$$BO \xrightarrow{\Delta} BO \times BO \xrightarrow{\sigma^r \times (\psi^r - 1)} BO_{\otimes} \times BO \xrightarrow{\omega \times \omega^{\perp}} W \times W^{\perp},$$

and

$$W \times W^{\perp} \xrightarrow{\nu \times \nu^{\perp}} BO \times BO_{\otimes} \xrightarrow{\rho^r \times (\psi^r/1)} BO_{\otimes} \times BO_{\otimes} \xrightarrow{\phi} BO_{\otimes}.$$

It is unusual to encounter pullbacks (as opposed to weak pullbacks, for which the uniqueness clause in the universal property is deleted) in the homotopy category. However, the equivalences of the theorem imply the following result.

Corollary 4.5. At p, the following diagram is a pullback in the homotopy category:

$$\begin{array}{ccc} BO & \xrightarrow{\psi^r - 1} & BSpin \\ \downarrow{\scriptstyle \sigma^r} & & \downarrow{\scriptstyle \rho^r} \\ BO_{\otimes} & \xrightarrow{\psi^r / 1} & BSpin_{\otimes} \end{array}$$

Write X_p for the localization at p of any space $X^{r(p)}$ which appears in the J-theory diagram and write f_p for the localization at p of $f^{r(p)}$ when f is one of the first five Greek letters. We thus have J_p, M_p, C_p, and BC_p.

Corollary 4.6. At p, the following composites are equivalences:

$$J_p \xrightarrow{\alpha_p} SF \xrightarrow{\varepsilon_p} J_{\otimes p} \quad \text{and} \quad M_p \xrightarrow{\beta_p} (SF;kO)/Spin \xrightarrow{\delta_p} M_p .$$

Proof. According to the J-theory diagram, these composites are maps of fibres induced from the pullback diagram of the previous corollary. It follows trivially that these composites induce monomorphisms, and therefore (by finiteness) isomorphisms, on homotopy groups.

Now elementary chases of the J-theory diagram yield the following interpretations of Adams' work, which are based on ideas and results of Sullivan [72; unpublished].

Theorem 4.7. At p = 2, the following composites are equivalences:

(i) $BC_2 \times BSpin \xrightarrow{\pi \times g} B(SF;kO) \times B(SF;kO) \xrightarrow{\phi} B(SF;kO)$

(ii) $C_2 \times Spin \xrightarrow{\Omega\pi \times \Omega g} (SF;kO) \times (SF;kO) \xrightarrow{\phi} (SF;kO)$

(iii) $C_2 \xrightarrow{\tau\Omega\pi} (SF;kO)/\,Spin$

(iv) $C_2 \times BO \xrightarrow{\pi\tau\Omega\pi \times \gamma_2} SF/Spin \times SF/Spin \xrightarrow{\phi} SF/Spin$

(v) $C_2 \times J_2 \xrightarrow{\Omega(q\pi)\times \alpha_2} SF \times SF \xrightarrow{\phi} SF$

Theorem 4.8. At $p > 2$, the following composites are equivalences:

(i) $BC_p \times W \times W^{\perp} \xrightarrow{1\times\nu\times\nu^{\perp}} BC_p \times BSpin \times BO_{\otimes} \xrightarrow{\pi\times g\times\tau} B(SF;kO)^3 \xrightarrow{\phi} B(SF;kO)$

(ii) $C_p \times \Omega W \times \Omega W^{\perp} \xrightarrow{1\times\Omega\nu\times\Omega\nu^{\perp}} C_p \times Spin \times SO_{\otimes} \xrightarrow{\Omega\pi\times\Omega g\times\Omega\tau} (SF;kO)^3 \xrightarrow{\phi} (SF;kO)$

(iii) $C_p \times M_p \xrightarrow{\tau\Omega\pi\times\beta_p} (SF;kO)/Spin \times (SF;kO)/Spin \xrightarrow{\phi} (SF;kO)/Spin$

(iv) $C_p \times BO \xrightarrow{\pi\tau\Omega\pi\times\gamma_p} SF/Spin \times SF/Spin \xrightarrow{\phi} SF/Spin$

(v) $C_p \times J_p \xrightarrow{\Omega(q\pi)\times\alpha_p} SF \times SF \xrightarrow{\phi} SF$

The behavior of $\tau: BO_{\otimes} \to B(SF;kO)$ and $g: BSpin \to B(SF;kO)$ on those parts of their domains which do not enter into the splitting of $B(SF;kO)$ is analyzed in the following immediate consequences of Lemma 4.3, Theorem 4.4, and the J-theory diagram. We agree to write ζ^{-1} for any chosen homotopy inverse to a homotopy equivalence ζ.

Corollary 4.9. At 2, $\tau: BO_{\otimes} \to B(SF;kO)$ is homotopic to the composite

$$BO_{\otimes} \xrightarrow{(\sigma^3)^{-1}} BO \xrightarrow{\psi^3-1} BSpin \xrightarrow{g} B(SF;kO).$$

Corollary 4.10. At $p > 2$, the composite $\tau\nu: W \to BO_{\otimes} \to B(SF;kO)$ is homotopic to both of the composites in the following diagram:

$$\begin{array}{ccccccccc}
W & \xrightarrow{(\omega\sigma^r\nu)^{-1}} & W & \xrightarrow{\nu} & BO & \xrightarrow{\psi^r-1} & BO & \xrightarrow{g} & B(SF;kO) \\
 & & \downarrow{\scriptstyle\nu} & & & & \uparrow{\scriptstyle\nu} & & \\
 & & BO & \xrightarrow{\psi^r-1} & BO & \xrightarrow{\omega} & W & &
\end{array}$$

Corollary 4.11. At $p > 2$, the composite $g\nu^{\perp}: W^{\perp} \to BO \to B(SF;kO)$ is homotopic to both of the composites in the following diagram:

$$\begin{array}{ccccccc} W^{\perp} & \xrightarrow{(\omega^{\perp}(\psi^r-1)\nu^{\perp})^{-1}} & W^{\perp} & \xrightarrow{\nu^{\perp}} BO \xrightarrow{\sigma^r} & BO & \xrightarrow{\tau} & B(SF;kO) \\ & & \downarrow \nu^{\perp} & & \uparrow \nu^{\perp} & & \\ & & BO & \xrightarrow{\sigma^r} BO_{\otimes} \xrightarrow{\omega^{\perp}} & W^{\perp} & & \end{array}$$

Here the lower routes are relevant to the splitting of $B(SF;kO)$, while the upper routes are more readily interpreted bundle theoretically.

Define global spaces $C = \times_p C_p$, $J = \times_p J_p$, etc. The spaces J and C are often called Im J and Coker J. J_2 is usually defined as the fibre of $\psi^3-1: BSO \to BSO$ localized at 2; this gives the same homotopy type as our J_2, but with a different H-space structure. In view of the key role played by g in the results above and in the study of the groups $JO(X)$, the present definition is preferable. It is also preferable on categorical grounds, as was noted by Tornehave [unpublished] and will be explained in VIII §3. In terms of stable stems, we choose to ignore the anomalous fact that η is in the image of J (anomalous because η^2 is not in the image of J), preferring instead to regard η as the first element of the periodic family not in the image of J in the $(8j+1)$-stems (see Remarks 5.3). Sullivan first defined the spaces C_p. Actually, his (unpublished) C_2 is the fibre of $f: F/O \to BSO_{\otimes}$ localized at 2. This definition is equivalent to ours by Theorem 4.7 and Lemma 3.4. The definition of BC_2 given here is new.

5. JSpin(X) and the δ and ε invariants

In this section and the next, the base spaces X of bundles are to be connected finite CW-complexes. To derive global bundle theoretic consequences from the local results of the previous section, we shall rely on the following basic fact in the theory of localization [23, V §6, 7 or 48, VII]. Let Y_0 denote the localization of a connected H-space Y at Q.

Theorem 5.1. For any set of primes T, the natural map

$$[X, Y_T] \to \underset{p \in T}{\text{pullback}} \; ([X, Y_p] \to [X, Y_0])$$

is an injection and is a bijection if $\pi_* Y$ is of finite type.

We shall generally be concerned with simply connected Y, when the brackets may be taken in the sense of unbased maps. When T is the set of all primes, $Y = Y_T$. The fact that $(BF)_0 \simeq *$ will allow us to ignore rational coherence below. We shall write f_p for the localization of a classifying map f at p and we shall use the same letter for an element of KG(X) and for its classifying map; we drop the curly brackets used earlier to distinguish stable from unstable bundles or oriented bundles.

We use Corollary 4.5 as a substitute for the main technical result of [4] in the following mixed local and global version of Adams' analysis [2, 3, 4] of the groups JO(X). Note that an F-trivial stable O-bundle admits a reduction to Spin (compare Lemma 3.4).

Theorem 5.2. The following are equivalent for a stable Spin-bundle ξ over X.

(i) ξ is trivial as a stable spherical fibration.

(ii) There exists a unit $\zeta \in KO(X)$ such that $\rho^r \xi = \psi^r \zeta / \zeta$ in $KSpin[1/r](X)$ for every integer $r \geq 2$.

(iii) For each prime p, there exists a unit $\zeta_p \in KO_p(X)$ such that $\rho^r \xi_p = \psi^r \zeta_p / \zeta_p$ in $KSpin_p(X)$, $r = r(p)$.

(iv) For each prime p, there exists an element $\eta_p \in KO_p(X)$ such that $\xi_p = \psi^r \eta_p - \eta_p$ in $KSpin_p(X)$, $r = r(p)$.

Proof. Regard ξ as a map $X \to BSpin$. If $Bj \circ \xi \simeq *$, there exists $\xi' : X \to SF/Spin$ such that $q\xi' \simeq \xi$. Define $\zeta = f\xi'$, $f: SF/Spin \to BO_{\otimes}$. Then (ii) holds by the J-theory diagram. (ii) trivially implies (iii) and (iii) implies (iv) by Corollary 4.5. If (iv) holds, $Bj \circ \xi$ is null homotopic because its localization at each prime is null homotopic.

Atiyah and Tall [15] gave a purely algebraic analog of the local equivalence (iii) $\Leftrightarrow$ (iv). Key points in their approach were the extension by continuity of the Adams operations ψ^r on p-adic γ-rings to p-adic integers r and use of the fact that r(p) generates the topological group of units of the p-adic integers (if $p > 2$). These algebraic considerations gained geometric content with Sullivan's introduction of the p-adic completions of spaces [72,73]. In particular, he pointed out that the p-adic completion of BO can be split by use of the obvious algebraic splitting [15, p. 284] of the functor to p-adic γ- rings obtained by p-adically completing $\widetilde{KO}(X)$ for X finite. Nevertheless, it does not seem to me that passage to p-adic completion would yield substantive additional information in the stable parts of Adams' and Sullivan's work discussed in this chapter.

We also prefer not to use the cocycle condition emphasized by Bott and Sullivan [22,72] or the periodicity condition emphasized by Adams [4]. These conditions make sense only when one considers the cannibalistic classes ρ^r for general values of r, rather than just for the r(p). These classes give no new information, and the cocycle and periodicity conditions can be viewed as formalizations of why they give no new information.

The following three results analyze the kernels of the natural transformations represented by the maps $q: B(SF;kO) \to BSF$, $g: BSpin \to B(SF;kO)$, and $\tau: BO_{\otimes} \to B(SF;kO)$. Recall that ψ^r was defined on the Abelian group $\widetilde{K}(SF;kO)[1/r](X)$ by its action on orientations. The difference $\psi^r - 1$ is represented by the map $B\psi^r/1$ on the classifying space $B(SF;kO)[1/r]$.

Theorem 5.3. Let (ξ, μ) be a kO-oriented stable spherical fibration over X and consider the following statements.

(i) The underlying stable spherical fibration ξ is trivial.

$(ii)_p$ For the prime p, there exists an element $(\eta_p, \nu_p) \in \widetilde{K}(SF;kO)_p(X)$ such that $(\xi_p, \mu_p) = \psi^r(\eta_p, \nu_p) - (\eta_p, \nu_p)$, $r = r(p)$.

Statement (i) implies statement $(ii)_p$ for each odd prime p and, provided that $(\psi^3 - 1)\widetilde{K}Spin_2(X) = (\psi^3 - 1)\widetilde{K}O_2(X) \subset \widetilde{K}Spin_2(X)$, for the prime 2. Conversely, statements $(ii)_p$ for all primes p imply statement (i).

Proof. Assume (i). Then (ξ, μ) is classified by $\tau_*(\zeta)$ for some map $\zeta: X \to BO_{\otimes}$. At $p = 2$, $\mathrm{Im}\, \tau_* \subset \mathrm{Im}\,(B\psi^3/1)_*$ under the stated proviso by Corollary 4.9 and Proposition 3.8; note that the proviso certainly holds if $H^2(X; Z_2) = 0$. At $p > 2$, $\mathrm{Im}(\tau\nu)_* \subset \mathrm{Im}(B\psi^r/1)_*$ by Corollary 4.10 and Proposition 3.8 while $\mathrm{Im}(\tau\nu^{\perp})_* \subset \mathrm{Im}(B\psi^r/1)_*$ by Lemma 4.3, Theorem 4.4 and Proposition 2.2. For the converse, $q(B\psi^r/1) \simeq *$ since $q \circ B\psi^r = q$ and therefore $(ii)_p$ for all p implies (i) by Theorem 5.1.

Proposition 5.4. Let ξ be a stable Spin-bundle over X. Then $(\xi, \mu(g))$ is trivial as a kO-oriented stable spherical fibration if and only if $\rho^{r(p)}\xi_p = 1 \in KSpin_p(X)$ for each prime p.

Proof. This is immediate from Theorems 4.4 and 5.1, Corollary 4.11, and a chase of the J-theory diagram.

Proposition 5.5. Let μ be a (special) unit in KO(X). Then (ε, μ) is trivial as a kO-oriented stable spherical fibration if and only if $\psi^{r(p)}\mu_p = \mu_p$ for all primes p.

Proof. This is immediate from Theorems 4.4 and 5.1, Corollaries 4.9 and 4.10, and a chase of the J-theory diagram.

The analysis in the rest of this section elaborates and makes rigorous a speculative program proposed by Adams [4, §7] and amplified by Sullivan [72, §6]. It will emerge that the theoretical framework envisioned by Adams leads to new bundle theoretic analogs δ and ε of the d and e invariants used in his computations of stable stems in [5]. Of course, the results of these computations are visible in the J-theory diagram.

Remarks 5.6. $\pi_* J_p$ can be read off from Theorem 3.5 and the homotopy exact sequence of the p-local fibration $J_p \to BO_p \xrightarrow{\psi^{r(p)}-1} BSpin_p$. For $i > 2$, the image of $J: \pi_i SO = \pi_i Spin \to \pi_i SF$ can then be read off from the splitting of SF at p. The map $e_*: \pi_* SF \to \pi_* BO_\otimes$ detects elements $\mu_i \in \pi_i SF$ of order 2, where $i \equiv 1$ or $2 \bmod 8$ and $i > 0$, such that μ_i comes from an element of $\pi_i J_2$ which is not in the image of $\pi_i Spin$.

Clearly e_* corresponds via adjunction to Adams' d-invariant (which assigns the induced homomorphism of reduced real K-theory to a map $S^{n+k} \to S^n$). Delooping the map e (which is an infinite loop map) and generalizing to arbitrary X, we can reinterpret this invariant as follows (compare IV.3.2).

Definition 5.7. For $\xi \in \widetilde{K}SF(X)$, define $\delta(\xi) = w(\xi; kO) \in BO^1_\otimes(X^+)$ to be the obstruction to the kO-orientability of ξ; equivalently, for $\xi \in [X^+, BSF]$, $\delta(\xi) = Be \circ \xi \in [X^+, BBO_\otimes]$.

Adams' e-invariant is defined on (a subgroup of) the kernel of d, and our ε-invariant will be defined on the kernel of δ. Of course, the latter kernel is just the group $Q(SF; kO)(X)$ of kO-orientable stable SF-bundles. Before defining ε, we note that Theorems 5.1 and 5.2, together with Lemmas 3.1 and 3.4, imply the following result.

Corollary 5.8. For G = Spin, SO, or O, $JG(X)$ is naturally isomorphic to

$$\times_p ([X^+, BG_p]/(\psi^{r(p)} - 1)_*[X^+, BO_p]) .$$

This suggests the following definition.

Definition 5.9. For G = Spin, SO, or O, define $JG_\otimes(X)$ to be the group

$$\times_p ([X^+, (BG_\otimes)_p]/(\psi^{r(p)}/1)_*[X^+, (BO_\otimes)_p]) .$$

Of course, the groups $JG(X)$ and $JG_\otimes(X)$ are abstractly isomorphic. In the case of Spin, the J-theory diagram yields a geometrically significant choice of isomorphism.

Definition 5.10. Define $\varepsilon : Q(SF;kO)(X) \to JSpin_\otimes(X)$ as follows. Given a kO-orientable stable SF-bundle ξ, choose a kO-orientation μ, localize at p, and apply the cannibalistic class $c(\psi^{r(p)})$. The image of this class in the p-component of $JSpin_\otimes(X)$ is independent of the choice of μ, and $\varepsilon(\xi) = \times_p c(\psi^{r(p)})(\xi, \mu)$. Equivalently, for $\xi : X \to BSF$ such that $Be \circ \xi \simeq *$, choose $\overline{\xi} : X \to B(SF;kO)$ such that $q\overline{\xi} \simeq \xi$. If also $q\overline{\xi}' \simeq \xi$, then $\overline{\xi}'/\overline{\xi} \simeq \tau\zeta$ for some $\zeta : X \to BO_\otimes$ and thus, at each prime p,

$$c(\psi^r)\overline{\xi}'_p/c(\psi^r)\overline{\xi}_p \simeq c(\psi^r)(\overline{\xi}'_p/\overline{\xi}_p) \simeq (\psi^r/1)\zeta_p , \qquad r = r(p).$$

Therefore $\varepsilon(\xi) = \times_p c(\psi^{r(p)})\overline{\xi}_p$ is a well-defined element of $JSpin_\otimes(X)$.

We need one more definition.

Definition 5.11. Let $C(X)$ denote the set of stable kO-oriented SF-bundles (ξ, μ) over X with local cannibalistic classes $c(\psi^{r(p)}) = 1 \in KSpin_p(X)$ for each prime p. Equivalently, if (ξ, μ) is classified by $\overline{\alpha}$, it is required that $c(\psi^{r(p)}) \circ \overline{\alpha}_p : X \to (BSpin_\otimes)_p$ be null homotopic for each prime p.

It is immediate from the form of the splittings of the $B(SF;kO)_p$ that $C(X)$ is classified by the space BC.

Theorem 5.12. The composite $JSpin(X) \subset Q(SF;kO)(X) \xrightarrow{\varepsilon} JSpin_{\otimes}(X)$ is an isomorphism, $C(X)$ maps monomorphically onto the kernel of ε under neglect of orientation, and therefore

$$Q(SF;kO)(X) = JSpin(X) \oplus C(X).$$

Proof. The first clause holds by comparison of Corollary 5.8 and Definition 5.9 with the equivalences of Theorem 4.4. The second clause holds since $(q\pi)_*: [X, BC] \to [X, BSF]$ is a monomorphism by the splitting of $B(SF;kO)_p$ and Corollaries 4.9 and 4.10 (which show that $\pi_*[X, BC_p]$ intersects $\tau_*[X, (BO_{\otimes})_p]$ trivially) and since $C(X)$ clearly maps onto the kernel of ε.

Comparison of Theorems 5.3 and 5.12 may be illuminating, particularly at the prime 2.

We discuss the relationship between Adams' e-invariant and our ε-invariant in the following remarks.

Remarks 5.13. Let $\varepsilon = \times_p \varepsilon_p : SF \to J_{\otimes} = \times_p J_{\otimes p}$. A straightforward chase of the J-theory diagram allows us to identify the ε-invariant on $\mathrm{Ker}(Be)_* \subset \pi_* BSF$ with the induced homomorphism

$$\varepsilon_*: \mathrm{Ker}(\pi_* SF \xrightarrow{e_*} \pi_* BO_{\otimes}) \to \mathrm{Ker}(\pi_* J_{\otimes} \to \pi_* J_{\otimes 2} \xrightarrow{\pi_*} \pi_* BO_2)$$

(the kernels being taken to avoid the elements μ_i of Remarks 5.6 and their images in the 2-component of $\pi_* J_{\otimes}$). On the other hand, an inspection of [5, §7 and §9] will convince the reader that Adams' real e-invariant (denoted e'_R or e_R in [5]) admits precisely the same description. Indeed, it can be seen in retrospect that Adams' calculation of the e-invariant on the image of J in [5, §10] amounts to a direct geometric comparison between the two invariants.

We complete this section with the development of a more conceptual description of the functor $C(X)$ than that given by Definition 5.11. We require some preliminaries.

Notations 5.14. Define bo, bso, and bspin to be the o-connected, 1-connected, and 2-connected covers of the spectrum kO; similarly, define bu and bsu to be the 0-connected and 2-connected covers of kU. In each case, the zeroth space is the one suggested by the notation. (Warning: bo and bu are usually taken as our kO and kU, this being a pointless waste of a useful notation.)

Lemma 5.15. Fix $r \geq 2$ and localize all spectra away from r. Then $\psi^r - 1: kO \to kO$ lifts uniquely to $\psi^r - 1: kO \to bspin$.

Proof. $\psi^r - 1$ obviously lifts uniquely to a map into bo. Recall that Bott periodicity implies that if $\eta: Q_\infty S^1 \to Q_\infty S^0$ is the unique nontrivial map, then

$$1 \wedge \eta: \Sigma KO = KO \wedge S^1 \cong KO \wedge Q_\infty S^1 \to KO \wedge Q_\infty S^0 \cong KO$$

is equivalent to the fibre of $c: KO \to KU$ (where KO and KU denote the periodic Bott spectra). Passage to associated connective spectra (II §2) yields a map $\bar{\eta}: \Sigma kO \to kO$ with the same behavior on homotopy groups in non-negative degrees as $1 \wedge \eta$, and $\bar{\eta}$ obviously lifts uniquely to a map $\bar{\eta}: \Sigma kO \to bo$. Its adjoint $\tilde{\eta}: kO \to \Omega bo$ clearly maps $\pi_0 kO$ onto $\pi_0 \Omega bo = Z_2$. Write H^* for cohomology with coefficients in Z_2 and recall that $H^* kO = A/ASq^1 + ASq^2$, where A denotes the mod 2 Steenrod algebra (e.g., by [7, p. 336]). In particular, $H^0 kO = Z_2$ and $H^1 kO = 0 = H^2 kO$. The fibration $bso \to bo \to K(Z_2, 1)$ gives an exact sequence

$$[kO, \Omega bo] \to H^0 kO \to [kO, bso] \to [kO, bo] \to 0$$

in which the first map is surjective by the properties of $\tilde{\eta}$. The fibration bspin $\to$ bso $\to$ $K(Z_2, 2)$ gives an isomorphism $[kO, bspin] \to [kO, bso]$ and the conclusion follows.

Definition 5.16. Define $j_p \in H\mathcal{J}$ to be the completion at p of the fibre of $\psi^{r(p)}-1\colon kO \to bspin$ and define $j = \times_p j_p$. Also, for use in VIII, define jO_2 to be the completion at 2 of the fibre of $\psi^3-1\colon kO \to bso$ and define JO_2 to be the fibre of $\psi^3-1\colon BO \to BSO$ at 2.

The use of completions is innocuous here (since the homotopy groups are finite in positive degrees) and serves to ensure that j is a ring spectrum. Indeed, we shall prove in VIII.3.2 that j_p is a ring spectrum such that the natural map $\kappa\colon j_p \to kO$ (completed at p) is a map of ring spectra.

Theorem 5.17. The spaces BC and B(SF; j) are equivalent. Therefore

$$C(X) = \widetilde{K}(SF; j)(X)$$

is the group of j-oriented stable spherical fibrations.

Proof. By II §2, the zeroth space of j_p is equivalent to $J_p \times \hat{Z}_{(p)}$; its 1-component is $J_{\otimes p}$, as an H-space, by VIII §3. The 1-component of the zeroth space of j is $J_\otimes = \times_p J_{\otimes p}$ and the projections give a homotopy commutative diagram

$$\begin{array}{ccccc} J_\otimes & \xrightarrow{\tau} & B(SF;j) & \xrightarrow{q} & BSF \\ \| & & \downarrow & & \downarrow \simeq \\ \times_p J_{\otimes p} & \xrightarrow{\times_p \tau} & \times_p B(SF;j_p)_p & \xrightarrow{\times_p q} & \times_p (BSF)_p \end{array}$$

We conclude that $B(SF;j)$ is equivalent to $\times_p B(SF;j_p)_p$; compare Lemma 2.4. Fix $r = r(p)$ and complete all spaces at p. A j_p-oriented stable spherical fibration clearly lies in $C(X)$ when regarded via $\kappa\colon j_p \to kO$ as a kO-oriented stable spherical fibration. Since the homotopy groups of $B(SF;j_p)$ are obviously finite, there are no $\varprojlim^1$ problems and we conclude that the composite

$$B(SF;j_p) \xrightarrow{B\kappa} B(SF;kO) \xrightarrow{c(\psi^r)} BSpin_{\otimes}$$

is null homotopic. There results a lift $\zeta: B(SF;j_p) \to BC_p$. Consider the following diagram, in which $(SF;j_p) = \Omega B(SF;j_p)$:

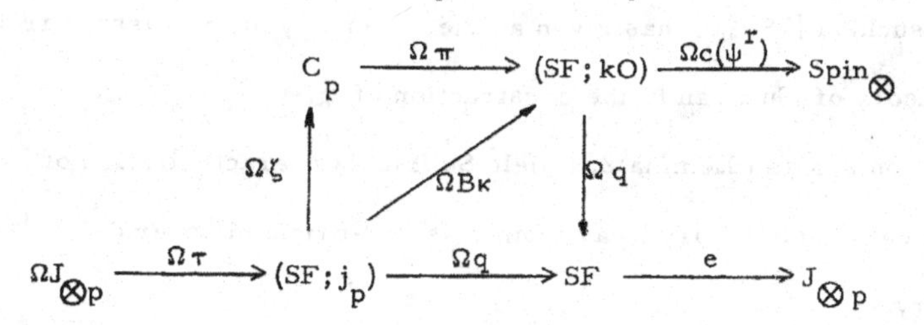

We have that $e_*: \pi_* SF \to \pi_* J_{\otimes p}$ is a split epimorphism by Corollary 4.6. By Theorems 4.7 and 4.8, $(\Omega q \Omega \pi)_*$ maps $\pi_* C_p$ monomorphically onto Ker e_*. In the bottom row, $(\Omega q)_*$ maps $\pi_*(SF;j_p)$ monomorphically onto Ker e_*. Therefore $(\Omega\zeta)_*$ is a monomorphism and thus an isomorphism. Delooping, we conclude that $\zeta_*: \pi_* B(SF;j_p) \to \pi_* BC_p$ is an isomorphism.

6. Sullivan's analysis of topological bundle theory

The following basic theorem is due to Sullivan.

Theorem 6.1. There exists a $kO[1/2]$-orientation $\bar{g}$ of STop. The localization away from 2 of the H-map $\bar{f}: F/Top \to BO_{\otimes}[1/2]$ associated to $\bar{g}: BSTop \to B(SF;kO[1/2])$ is an equivalence.

The first statement is proven in [72, §6]. Actually, the proof that $\bar{g}$ is an H-map is omitted there.[1] It is easy to see that $\bar{g}$ is multiplicative modulo torsion, however, and this suffices for the discussion of $\bar{f}$ as an

1. A proof will be given in Theorem 7.16 below.

H-map. Note that, by Lemma 2.5, $\overline{f}$ is the unique H-map such that the square

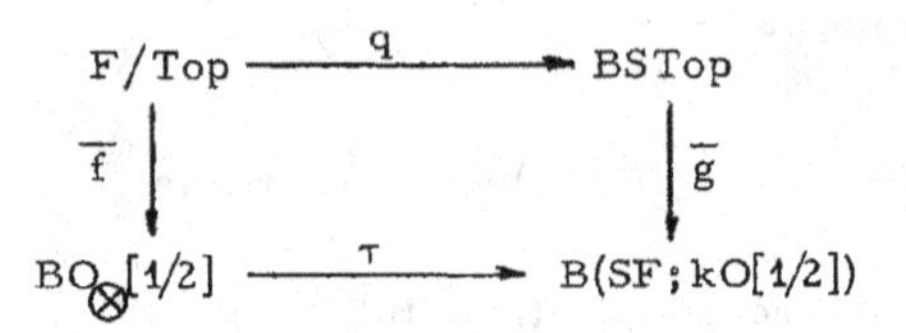

is homotopy commutative. The second statement of the theorem is proven in [71], and Tsuchiya [78, §6] has given a brief summary of the passage from this earlier theory of Sullivan to the construction of $\overline{g}$.

Proposition 2.3 and Lemma 2.4 yield Sullivan's characterization of stable topological bundle theory away from 2 as kO-oriented spherical fibration theory.

Corollary 6.2. $\overline{g}$: BSTop[1/2] → B(SF; kO)[1/2] is an equivalence.

In the analog of the J-theory diagram on the following page, the solid arrow diagram exists globally and the entire diagram exists and is homotopy commutative away from 2 and r. θ^r is defined to be $c(\psi^r)\overline{g}Bi$, and Sullivan [72, 6.81-6.82] obtained the following calculation by comparing the cannibalistic classes θ^r and ρ^r; explicitly, $\theta^r(x) = \rho^r(\psi^2(2x) - \psi^4(x))$.

Theorem 6.3. Let $x \in \pi_{4j}BO[1/2, 1/r]$, $j > 0$. Then

$$\theta^r x = 1 + 2^{2j}(r^{2j} - 1)(1 - 2^{2j-1})\alpha_{2j} x .$$

λ^r is defined to be $Bi \cdot \gamma^r$. By Remarks 3.8, $\overline{f}\lambda^r$ is homotopic to θ^r and the fibres of λ^r and of θ^r are thus equivalent; by abuse, we denote both by N^r. The maps $\overline{\beta}^r$ and $\overline{\delta}^r$ are obtained by Barratt-Puppe sequence arguments, and the remaining maps already appear in the J-theory diagram. Proposition 2.3 and the diagram imply the following analog of Proposition 3.8.

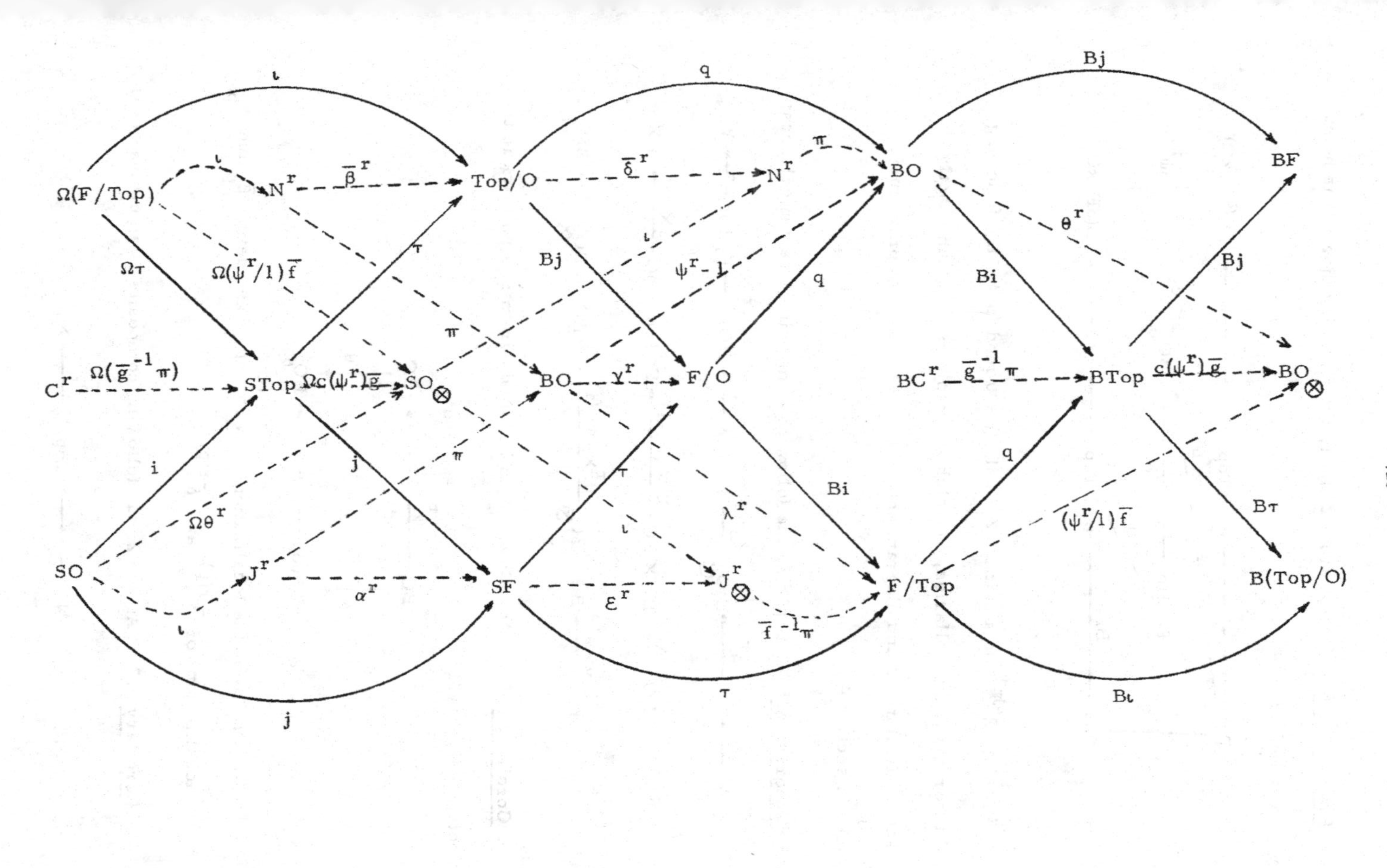
Ω(F/Top)
N^r
Top/O
N^r
BO
BF
C^r
STop
SO⊗
BO
F/O
BC^r
BTop
BO⊗
SO
J^r
SF
J^r⊗
F/Top
B(Top/O)
Ωτ
Ω(ψ^r/1)f̄
Ω(ḡ^-1 π)
Ωc(ψ^r)ḡ
Ωθ^r
β̄^r
δ̄^r
γ^r
θ^r
ψ^r-1
λ^r
α^r
ε^r
f̄^-1 π
c(ψ^r)ḡ
ḡ^-1 π
(ψ^r/1)f̄
Bj
Bi
Bτ
Bι

Proposition 6.4. Away from 2 and r, the following diagram is homotopy commutative:

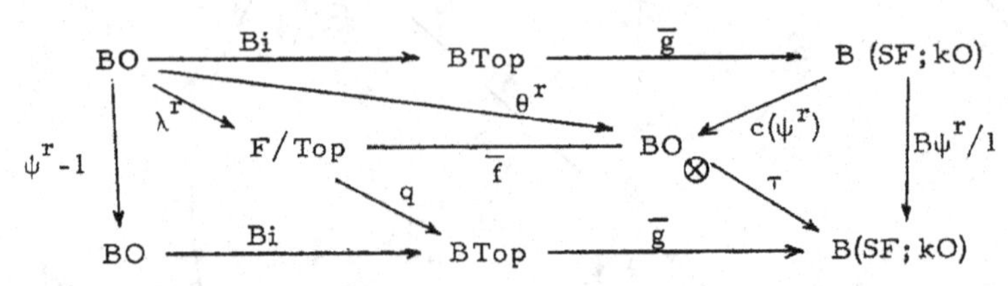

Since $1 - 2^{2j-1}$ is a unit in $Z_{(p)}$ if $2j \equiv 0 \bmod (p-1)$, the following local theorems and corollaries, in which $r = r(p)$, result from exactly the same calculations and diagram chases that were used to prove their analogs in section 4.

Theorem 6.5. At $p > 2$, the following composites are equivalences:

$$W \xrightarrow{\nu} BO \xrightarrow{\theta^r} BO_{\otimes} \xrightarrow{\omega} W, \quad W \xrightarrow{\nu} BO \xrightarrow{\lambda^r} F/Top \xrightarrow{\omega} W,$$

$$BO \xrightarrow{\Delta} BO \times BO \xrightarrow{\lambda^r \times (\psi^r - 1)} F/Top \times BO \xrightarrow{\omega \times \omega^{\perp}} W \times W^{\perp}$$

and

$$W \times W^{\perp} \xrightarrow{\nu \times \nu^{\perp}} BO \times BO_{\otimes} \xrightarrow{\theta^r \times (\psi^r/1)\overline{f}} BO_{\otimes} \times BO_{\otimes} \xrightarrow{\phi} BO_{\otimes}.$$

Corollary 6.6. At $p > 2$, the following diagram is a pullback in the homotopy category:

$$\begin{array}{ccc} BO & \xrightarrow{\psi^r - 1} & BO \\ \downarrow{\scriptstyle \lambda^r} & & \downarrow{\scriptstyle \theta^r} \\ F/Top & \xrightarrow{(\psi^r/1)\overline{f}} & BO_{\otimes} \end{array}$$

Define N_p to be the localization of $N^{r(p)}$ at p and define $\overline{\beta}_p$ and $\overline{\delta}_p$ to be the localizations of $\overline{\beta}^{r(p)}$ and $\overline{\delta}^{r(p)}$ at p.

Corollary 6.7. At $p > 2$, the following composite is an equivalence:

$$N_p \xrightarrow{\overline{\beta}_p} Top/O \xrightarrow{\overline{\delta}_p} N_p .$$

Theorem 6.8. At $p > 2$, the following composites are equivalences.

(i) $BC_p \times W \times W^{\perp} \xrightarrow{1 \times \nu \times \nu^{\perp}} BC_p \times BO \times F/Top \xrightarrow{\bar{g}^{-1}\pi \times Bi \times q} (BTop)^3 \xrightarrow{\phi} BTop.$

(ii) $C_p \times \Omega W \times \Omega W^{\perp} \xrightarrow{1 \times \Omega\nu \times \Omega\nu^{\perp}} C_p \times SO \times \Omega(F/Top) \xrightarrow{\Omega(\bar{g}^{-1}\pi) \times i \times \Omega q} (STop)^3 \xrightarrow{\phi} Stop.$

(iii) $C_p \times N_p \xrightarrow{\tau\Omega(\bar{g}^{-1}\pi) \times \bar{\beta}_p} Top/O \times Top/O \xrightarrow{\phi} Top/O.$

The odd primary parts of Brumfiel's calculations [24] of $\pi_* BTop$ and $\pi_* Top/O$ can be read off from the theorem and the diagram.

Corollary 6.9. At $p > 2$, the composite $q\nu : W \to F/Top \to BTop$ is homotopic to both of the composites in the following diagram:

$$\begin{array}{ccccccccc}
W & \xrightarrow{(\omega\lambda^r \nu)^{-1}} & W & \xrightarrow{\nu} & BO & \xrightarrow{\psi^r - 1} & BO & \xrightarrow{Bi} & BTop \\
 & & \downarrow \nu & & & & \uparrow \nu & & \\
 & & BO & \xrightarrow{\psi^r - 1} & BO & \xrightarrow{\omega} & W & &
\end{array}$$

Corollary 6.10. At $p > 2$, the composite $Bi \circ \nu^{\perp} : W^{\perp} \to BO \to BTop$ is homotopic to both of the composites in the following diagram:

$$\begin{array}{ccccccccc}
W^{\perp} & \xrightarrow{(\omega^{\perp}(\psi^r - 1)\nu^{\perp})^{-1}} & W^{\perp} & \xrightarrow{\nu^{\perp}} & BO & \xrightarrow{\lambda^r} & F/Top & \xrightarrow{q} & BTop \\
 & & \downarrow \nu^{\perp} & & & & \uparrow \nu^{\perp} & & \\
 & & BO & \xrightarrow{\lambda^r} & F/Top & \xrightarrow{\omega^{\perp}} & W^{\perp} & &
\end{array}$$

The bundle theoretic interpretations of the results above are evident from the diagram and the arguments of the previous section. Consider all bundle theories in sight as localized away from 2. Corollary 6.2 asserts that every kO[1/2]-oriented stable F-bundle has the form $(\xi, \mu(\bar{g}))$ for some Top-bundle ξ and that two stable Top-bundles ξ and ξ' are equal if $(\xi, \mu(\bar{g}))$ and $(\xi', \mu(\bar{g}))$ are equal as kO[1/2]-oriented stable F-bundles. (Here, away from 2, we may write F and Top but think in terms of the integrally oriented case.)

The Adams operation ψ^r acts on $\widetilde{K}Top(X)$ via its action on the Sullivan orientation (away from 2 and r). More precisely, the action of ψ^r on $\widetilde{K}(SF; kO)(X)$ is to be transported to $\widetilde{K}Top(X)$ along the equivalence $\overline{g}_*$. Similarly, the action of ψ^r on the group of (special) units $[X^+, BO_\otimes]$ in $KO(X)$ may be transported to $[X^+, F/Top]$ along $\overline{f}_*$. Then, with ψ^r acting trivially on $\widetilde{K}F(X)$, the transformations induced by the maps $Bj: BTop \rightarrow BF$, $Bi: BO \rightarrow BTop$, and $q: F/Top \rightarrow BTop$ commute with the ψ^r (by Propositions 6.4, 2.2, and 2.3 in the last two cases). The following three results analyze the kernels of these transformations. Again, the proofs are the same as for the analogous results of Section 5.

Theorem 6.11. Away from 2, the following are equivalent for a stable Top-bundle ξ over X.

(i) ξ is trivial as a stable spherical fibration.

(ii) For each odd prime p, there exists an element $\eta_p \in \widetilde{K}Top_p(X)$ such that $\xi_p = \psi^r \eta_p - \eta_p$, $r = r(p)$.

Proposition 6.12. Let ξ be a stable O-bundle over X. Then, away from 2, ξ is trivial as a stable Top-bundle if and only if $\theta^{r(p)}\xi_p = 1 \in KO_p(X)$ for each odd prime p.

Proposition 6.13. Let (ξ, τ) be an F-trivialized stable Top-bundle over X. Then, away from 2, ξ is trivial as a stable Top-bundle if and only if $\psi^{r(p)}(\xi, \tau) = (\xi, \tau)$ for each odd prime p (or, equivalently, $\psi^{r(p)}\zeta_p = \zeta_p$ for each odd prime p, where ζ is that unit of $KO[1/2](X)$ such that the Sullivan orientation of ξ is the cup product of ζ and the orientation induced by τ from the canonical orientation of the trivial stable F-bundle).

Away from 2, the δ-invariant may be interpreted as the obstruction to the existence of a topological structure (that is, a reduction of the structural monoid to Top) on a stable F-bundle. $Q(SF; kO)(X)$ may be interpreted as $JTop(X)$, and the ε-invariant may thus be regarded as defined on $JTop(X)$. $C(X)$ may be interpreted as the set of stable Top-bundles over X all of whose local cannibalistic classes $c(\psi^{r(p)})\bar{g}_p$ are trivial. Theorem 5.12 may then be interpreted as follows.

Theorem 6.14. Away from 2, the composite $JO(X) \subset JTop(X) \xrightarrow{\varepsilon} JO_{\otimes}(X)$ is an isomorphism, $C(X)$ maps monomorphically onto the kernel of ε under passage to fibre homotopy equivalence, and therefore

$$JTop(X) = JO(X) \oplus C(X).$$

Remark 6.15. For what it is worth, we note that there is a precise analog to Theorem 6.14 in which $B(Top/O)$ plays the role of BF. The proof is slightly more complicated, because rational coherence must be taken into account, but the conclusion is again that, away from 2, the image of

$$(B\iota)_*: [X^+, F/Top] \to [X^+, B(Top/O)]$$

is a direct summand of the image of

$$(B\tau)_*: [X^+, BTop] \to [X^+, B(Top/O)]$$

with complementary summand $C(X)$. This remark also has an analog in the J-theory case since the results of the next section imply that the J-theory diagram admits a lower right-hand corner.

Remark 6.16. We have used Top and F instead of PL and G since the former theories fit naturally into our general context. Stably and away from 2, there is no distinction. Unstably, Sullivan's $\bar{g}$ is the limit of $kO[1/2]$-orientations of $SPL(n)$-block bundles $\bar{g}(n): BSPL(n) \to B(SG(n); kO[1/2])$, where the classifying space on the right can be constructed either by the methods of [47] or by use of Brown's theorem. Haefliger and Wall's result [32] that $G(n)/PL(n) \to G/PL$

is an equivalence for $n \geq 3$, together with an unstable comparison diagram obtained by Barratt-Puppe sequence arguments and use of Lemmas 2.6 and 2.7, show that $\overline{g}(n)[1/2]$ is an equivalence for every $n \geq 3$. Note, however, that the block bundle version of $PL(n)$ used in this remark is not the one most relevant to geometric work in piecewise-linear topology.

§7. Infinite loop analysis of the main diagrams

In order to determine which of the various splittings we have obtained are actually splittings of infinite loop spaces, we must determine which maps in our main diagrams are infinite loop maps. It turns out that homotopy theoretic arguments, which can be thought of as ultimately based on how tightly Bott periodicity ties together the p-local k-invariants of BO and BU, coupled with representation theoretical calculations, yield a great deal of information about this question. The relevant arguments are due to Madsen, Snaith, and Tornehave [42] and will be outlined here. To begin with, these authors have proven the following analog for maps of the Adams-Priddy unique deloopability of spaces result, Theorem 4.2. Their proof is based on the fact that $[K(\hat{Z}_p, n), \widehat{BU}_p] = 0$ for $n \geq 3$. An alternative proof based on the techniques of Adams and Priddy [8] is possible. Let T be any set of primes. In all of the results of this section, our Theorems II.2.13 and II.2.14 show that the result for localizations at T follows immediately from the result for completions at p for $p \in T$.

Theorem 7.1. Let D and E be T-local or T-complete connective spectra the zeroth spaces of which have completions at each $p \in T$ equivalent to those of either BU or BSO. Then the natural homomorphism $[D,E] \to [D_0,E_0]$ is a monomorphism.

The proof of Madsen, Snaith, and Tornehave (MST henceforward) equally well yields the following complement except in the real case at $p = 2$ where the result is due to Ligaard [38].

Theorem 7.2. Let D and E be T-local or T-complete connective spectra the zeroth spaces of which have completions at each $p \in T$ equivalent, respectively, to those of U and BU or SO and BO or Spin and BSO. Then $[D, E] = 0$.

MST then proceed to an analysis of which H-maps $f: \widehat{BU}_p \to \widehat{BU}_p$ or $f: \widehat{BSO}_p \to \widehat{BSO}_p$ are in fact infinite loop maps. Such a map can be uniquely written in the form $f = f_1 + f_2\psi^p$, where f_1 and f_2 are H-maps and f_1 is prime to ψ^p (in a suitable sense), and their basic observation is that f is an infinite loop map if and only if $f_2 = 0$ (since f is then essentially a linear combination of the ψ^r with r prime to p). This assertion has the following consequence.

Theorem 7.3. Let X, Y, and Z be T-local or T-complete infinite loop spaces whose completions at each $p \in T$ are equivalent to those of BSO. Let $f: X \to Y$ and $g: Y \to Z$ be H-maps such that gf is an infinite loop map and either f or g is both an infinite loop map and a rational equivalence. Then the remaining map g or f is an infinite loop map.

Proof. By II.2.13, II.2.14, and Theorem 4.2, it suffices to show this when X, Y, and Z are $\widehat{BSO}_p$ as infinite loop spaces. For definiteness, let f be an infinite loop map and a rational equivalence. Then $f = f_1$, $g = g_1 + g_2\psi^p$, and $fg = fg_1 + fg_2\psi^p = fg_1$. But $fg_2 = 0$ implies $g_2 = 0$ by Lemma 2.8.

The criterion above for determining whether an H-map $\widehat{BSO}_p \to \widehat{BSO}_p$ is an infinite loop map will be interpreted representation theoretically at the end of VIII §1. MST prove the following pair of results by representation theoretical calculations based on [3] for ρ^r; the result for $\theta^r \simeq \rho^r(\psi^2 \circ 2 + \psi^4)$ follows.

<u>Lemma 7.4.</u> $\rho^r: BSO[1/r] \to BSO_{\otimes}[1/r]$ is an infinite loop map.

<u>Lemma 7.5.</u> $\theta^r: BO[1/2, 1/r] \to BO_{\otimes}[1/2, 1/r]$ is an infinite loop map.

The following analog is simpler, requiring only Theorem 4.2.

<u>Lemma 7.6.</u> $\psi^r: BSO_{\otimes}[1/r] \to BSO_{\otimes}[1/r]$ is an infinite loop map.

<u>Proof.</u> At p prime to r, Theorem 4.2 gives an equivalence $\xi: BSO \to BSO_{\otimes}$ of infinite loop spaces. By the argument in Remarks 3.7, ψ^r is homotopic to the infinite loop map $\xi\psi^r\xi^{-1}$.

At this point, it is convenient to insert a remark relevant only at the prime 2.

<u>Remark 7.7.</u> Recall from Anderson and Hodgkin [12] that $\widetilde{KO}^*(K(\pi, n)) = 0$ for $n \geq 2$ and all finite Abelian groups π. By use of II.3.2, II.2.10, and the non-splittable fibrations

$$bso \to bo \to \mathcal{K}(Z_2, 1) \quad \text{and} \quad bspin \to bso \to \mathcal{K}(Z_2, 2),$$

it is easy to deduce that

$$[\mathcal{K}(Z_2, n), kO] = 0 \text{ for } n \geq 0, \quad [\mathcal{K}(Z_2, n), bo] = 0 \text{ for } n \geq 0,$$

$$[\mathcal{K}(Z_2, n), bso] = 0 \text{ for } n \geq 1, \quad [\mathcal{K}(Z_2, n), bspin] = 0 \text{ for } n \geq 2,$$

$$[\mathcal{K}(Z_2, 0), bso] = Z_2 \quad \text{and} \quad [\mathcal{K}(Z_2, 1), bspin] = Z_2.$$

In the last two cases, both maps of spectra induce the trivial map $K(Z_2, 0) \to BSO$ and $K(Z_2, 1) \to BSpin$ on zeroth spaces.

In VIII §3, we shall prove that $c(\psi^r): B(SF; kO) \to BSpin_{\otimes}$ is an infinite loop map at p, where $r = r(p)$. Thus the fibre BC_p of $c(\psi^r)$ and its loop space C_p are infinite loop spaces. Define $\psi^r/1 = c(\psi^r)\tau: BO_{\otimes} \to BSpin_{\otimes}$ as an infinite loop map at p. On $BSO_{\otimes}$, this delooping is the one coming from Lemma 7.6 in view of Theorem 7.1. When $p = 2$, this definition fixes a choice of delooping of the trivial map $BO(1) \to BSpin_{\otimes}$, and in fact gives the non-trivial delooping. Define $J_{\otimes p}$ as an infinite loop space to be the fibre of $\psi^r/1$. We shall construct an infinite loop fibration $C_p \to SF \to J_{\otimes p}$ in VIII §3 and will show that it splits when $p > 2$ in VIII §4. By the following basic result of Hodgkin and Snaith [33; 70, §9], this shows that, to the eyes of K-theory, SF is equivalent to $J_{\otimes p}$.

Theorem 7.8. $\widetilde{K}^*(C_p) = 0$ and $\widetilde{KO}^*(C_p) = 0$; there are no non-trivial maps $C_p \to BSO$ on either the space or the spectrum level.

We can now prove the following analog of Theorems 7.1 and 7.2, which is due to MST at $p > 2$ and to Ligaard at $p = 2$. Their proofs are somewhat more difficult, but give more precise information.

Theorem 7.9. Let X be a T-local or T-complete infinite loop space whose completion at each $p \in T$ is equivalent to that of $BO_{\otimes}$. Then an H-map $f: J_{\otimes} \to X$ or $g: SF \to X$ is the zeroth map of at most one map of spectra.

Proof. We may work at p, with X replaced by BSO, in view of II. 2.13, II. 2.14, Lemma 3.1, and Theorem 4.2. Clearly the result for g will follow immediately from the result for f. With $r = r(p)$, consider the following diagram, the rows of which are infinite loop fibrations:

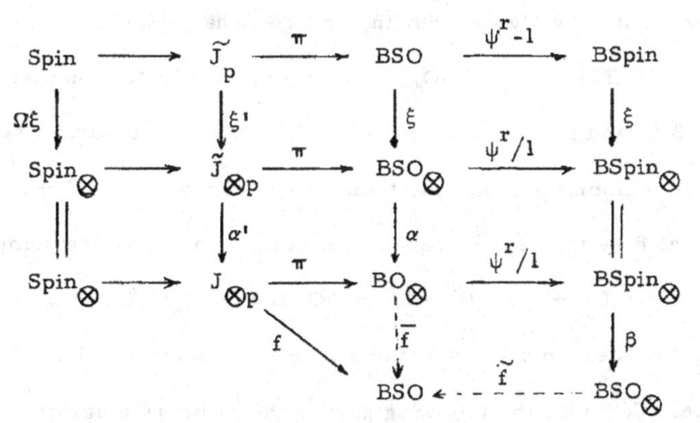

$$\begin{array}{ccccccc}
Spin & \longrightarrow & \widetilde{J}_p & \xrightarrow{\pi} & BSO & \xrightarrow{\psi^r - 1} & BSpin \\
\downarrow \Omega\xi & & \downarrow \xi' & & \downarrow \xi & & \downarrow \xi \\
Spin_{\otimes} & \longrightarrow & \widetilde{J}_{\otimes p} & \xrightarrow{\pi} & BSO_{\otimes} & \xrightarrow{\psi^r/1} & BSpin_{\otimes} \\
\| & & \downarrow \alpha' & & \downarrow \alpha & & \| \\
Spin_{\otimes} & \longrightarrow & J_{\otimes p} & \xrightarrow{\pi} & BO_{\otimes} & \xrightarrow{\psi^r/1} & BSpin_{\otimes} \\
 & & & \searrow f & \downarrow \bar{f} & & \downarrow \beta \\
 & & & & BSO & \xleftarrow{\tilde{f}} & BSO_{\otimes}
\end{array}$$

The maps ξ are infinite loop equivalences coming from Theorem 4.2, α, β, and the π are the natural infinite loop maps, and ξ' and α' are infinite loop maps coming from Barratt-Puppe sequence arguments (e.g. [48, I]). Clearly ξ' is an equivalence. By Hodgkin and Snaith [33, 4.7], the top fibration yields an exact sequence

$$0 \to PKO(BSpin) \xrightarrow{(\psi^r - 1)^*} PKO(BSO) \xrightarrow{\pi^*} PKO(\tilde{J}_p) \to 0$$

of primitive elements in p-complete K-theory. (They deal with KU, but the result for KO follows.) Let $f: J_{\otimes p} \to BSO$ be an infinite loop map which is trivial as a map of spaces. We must show that f is trivial as an infinite loop map. By Theorem 7.2, $f = \bar{f}\pi$ as an infinite loop map for some infinite loop map $\bar{f}: BO_{\otimes} \to BSO$ (because $[spin, bso] = 0$ and fibrations are negatives of cofibrations in the stable category [48, XI]). By the exact sequence of primitive elements (i.e., H-maps) and, when $p = 2$, Lemma 2.10, there is an H-map $\tilde{f}: BSO_{\otimes} \to BSO$ such that $\tilde{f}\beta\xi(\psi^r - 1) \simeq \bar{f}\alpha\xi$. By Theorem 7.3, $\tilde{f}$ is an infinite loop map. By Theorem 7.1, we conclude that $f\alpha'\xi' = \tilde{f}\beta\xi(\psi^r - 1)\pi$ is the trivial infinite loop map. Since $(\alpha')^*: [j_{\otimes 2}, bso] \to [\tilde{j}_{\otimes 2}, bso]$ is a monomorphism by Remark 7.7, f is also the trivial infinite loop map.

These results allow infinite loop analysis of the comparison diagram parts of the J-theory diagram and of its analog for topological bundle theory. The following result was noted by Madsen, Snaith, and Tornehave [42].

Proposition 7.10. $f: SF/Spin \to BSO_{\otimes}$ is (globally) an infinite loop map.

Proof. By II.2.13 and II.2.14, it suffices to prove the result with all spaces completed at p. By Lemmas 3.1 and 3.4, it suffices to consider $f: F/O \to BSO_{\otimes}$ (even at $p = 2$). Let B be the zeroth space of the cofibre in $H\mathscr{S}$ of the composite infinite loop map $\iota: C_p \to (SF; kO) \to SF \to F/O$ and let $\zeta_p: F/O \to BO$ be the natural map. By Theorems 4.7, 4.8, and 4.2, B is equivalent to BSO as an infinite loop space. Consider the following part of the J-theory diagram with $r = r(p)$

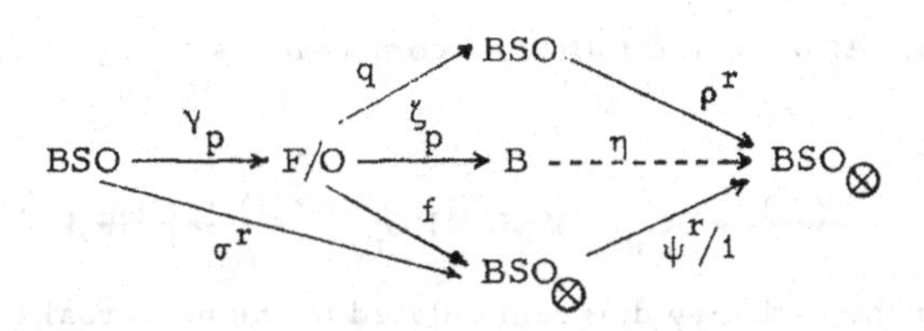

By Theorem 4.2 and Theorem 7.8, $\rho^r q\iota : C_p \to BSO_\otimes$ is the trivial infinite loop map, hence there is an infinite loop map $\eta: B \to BSO_\otimes$ such that $\eta\zeta_p = \rho^r q$ as infinite loop maps. By Remarks 3.7, $\sigma^r \simeq \rho^r$. Thus $\eta(\zeta_p\gamma_p) \simeq (\psi^r/1)\rho^r$. By Theorem 7.3 and Lemmas 7.4 and 7.6, $\zeta_p\gamma_p$ is an infinite loop map. Clearly $\zeta_p\gamma_p$ is a homotopy equivalence, and $f\gamma_p \simeq \rho^r(\zeta_p\gamma_p)^{-1}\zeta_p\gamma_p$ while f and $\rho^r(\zeta_1\gamma_1)^{-1}\zeta_p$ both restrict to the trivial map on C_p. It follows from the splitting of F/O in Theorems 4.7 and 4.8 that f is homotopic to the infinite loop map $\rho^r(\zeta_p\gamma_p)^{-1}\zeta_p$.

Theorem 7.11. The following is (globally) a commutative diagram of infinite loop spaces and maps:

$$\begin{array}{ccccccc}
SF & \xrightarrow{\tau} & SF/Spin & \xrightarrow{q} & BSpin & \xrightarrow{Bj} & BSF \\
\downarrow \chi & & \downarrow f & & \downarrow g & & \| \\
SF & \xrightarrow{e} & BO_\otimes & \xrightarrow{\tau} & B(SF; kO) & \xrightarrow{q} & BSF
\end{array}$$

Proof. By Theorem 7.9 and the previous proposition, the left square is a commutative diagram of infinite loop spaces and maps. As pointed out in section 2, a Barratt-Puppe sequence argument on the spectrum level gives an infinite loop map g': BSpin $\to$ B(SF; kO) which makes the right two squares commute on the infinite loop space level. On the space level, $g - g' = \tau h$ for some h: BSpin $\to BO_\otimes$. However, commutation of the middle square implies that $g_* = (g')_*$ on rational homology and thus that $h_* = 0$. Therefore $h \simeq 0$ by Lemmas 2.8, 2.10, and 3.1 and Theorem 4.2. Thus $g \simeq g'$.

Corollary 7.12. At p = 2, the following composite is an equivalence of infinite loop spaces:

$$BC_2 \times BSpin \xrightarrow{\pi \times g} B(SF; kO) \times B(SF; kO) \xrightarrow{\phi} B(SF; kO).$$

Corollary 7.13. At $p>2$, the following composite is an equivalence of infinite loop spaces:

$$BC_p \times W \times W^{\perp} \xrightarrow{1\times\nu\times\nu^{\perp}} BC_p \times BSpin \times BO_{\otimes} \xrightarrow{\pi\times g\times\tau} B(SF;kO)^3 \xrightarrow{\phi} B(SF;kO).$$

Those parts of the J-theory diagram related to the universal cannibalistic class $c(\psi^r)$ will be analyzed on the infinite loop level in VIII §3.

All remaining parts of the J-theory diagram depend on the Adams conjecture and thus on $\gamma^r: BO \to SF/Spin$. Madsen [41] has shown that γ^3 cannot be so chosen that its localization at 2, or that of α^3, is even an H-map. (See also [26, II.12.2]). Nevertheless, it seems likely that, away from 2 and r, γ^r can be chosen as an infinite loop map. The following conjecture is even a bit stronger.

Conjecture 7.14[1]. The complex Adams conjecture holds on the infinite loop space level. That is, for each r, the composite

$$BU \xrightarrow{\psi^r - 1} BU \xrightarrow{Bj} BSF$$

is trivial as an infinite loop map when localized away from r.

By II. 2.13, it suffices to work one prime at a time. The proof in VIII §4 that SF splits as $J_p \times C_p$ as an infinite loop space at each odd prime p will give explicit splitting maps $J_p \to SF$, but it is not known whether or not these maps are homotopic to (some choices of) α_p in the J-theory diagram.

Turning to the analysis of BTop away from 2, we have the following analog of Proposition 7.10, which was also noted by Madsen, Snaith, and Tornehave [42].

Proposition 7.15. $\overline{f}: F/Top[1/2] \to BO_{\otimes}[1/2]$ is an infinite loop map.

---

1. See the discussion following Remarks VIII. 4. 6.

Proof. Again, by II. 2.13 and II. 2.14, we may work on the p-complete level, $p > 2$. With $\zeta_p : F/O \to B$ as in the proof of Proposition 7.10, consider the following part of the main diagram in section 6, $r = r(p)$:

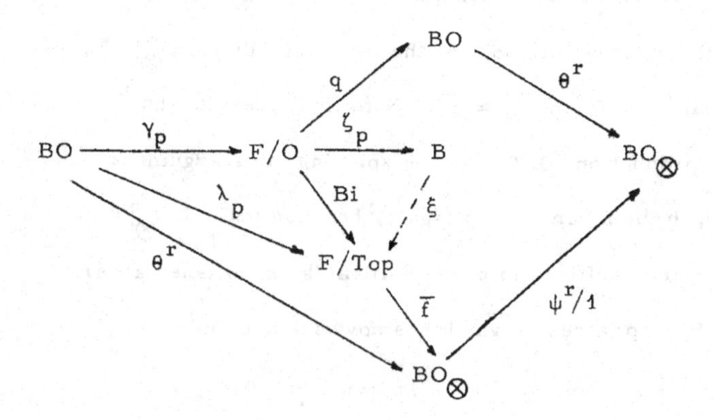

By Theorem 4.2 and Theorem 7.8, $Bi \circ \iota : C_p \to F/Top$ is the trivial infinite loop map, hence there is an infinite loop map $\xi : B \to F/Top$ such that $\xi\zeta_p = Bi$ as infinite loop maps. Now $\lambda_p \simeq \xi \circ (\zeta_p \gamma_p)$ is an infinite loop map, since $\zeta_p \gamma_p$ is an infinite loop map by the proof of Proposition 7.10, and thus $\overline{f}$ is an infinite loop map by Lemma 7.5 and Theorem 7.3.

Theorem 7.16. Away from 2, the following is a commutative diagram of infinite loop spaces and maps:

$$\begin{array}{ccccccc} SF & \xrightarrow{\tau} & F/Top & \xrightarrow{q} & BSTop & \xrightarrow{Bj} & BSF \\ \downarrow \chi & & \downarrow \overline{f} & & \downarrow \overline{g} & & \| \\ SF & \xrightarrow{e} & BO_{\otimes} & \xrightarrow{\tau} & B(SF; kO) & \xrightarrow{q} & BSF \end{array}$$

Therefore $\overline{f}$ and $\overline{g}$ are equivalences of infinite loop spaces.

Proof. Again, the left square commutes on the infinite loop level by Theorem 7.9, hence Barratt-Puppe sequence arguments give a map $\overline{g}'$: BSTop $\rightarrow$ B(SF; kO) which makes the right two squares commute on the infinite loop space level. $\overline{g} - \overline{g}' = \tau h$ for some map h: BSTop $\rightarrow BO_{\otimes}$, and the rationalization of h is null homotopic. It follows by use of the splitting of BSTop, the fact that $[BC_p, BO_{\otimes}] = 0$, and Lemma 2.8 that $h \simeq 0$ and $\overline{g} \simeq \overline{g}'$. Note that, despite the role played by the product on BSTop in its splitting, this argument does not depend on h being an H-map (again, because $[BC_p, BO_{\otimes}] = 0$) and therefore more than suffices to prove Sullivan's unpublished assertion that $\overline{g}$ is an H-map (a result which has nowhere been used in our work above).

By Corollary 7.13 and Theorem 7.16, BSTop splits at p as $BC_p \times W \times W^{\perp}$ as an infinite loop space. This fact, together with the firm grasp on BC_p as an infinite loop space given by VIII §3, has been used to obtain precise information on the characteristic classes (at $p > 2$) for stable topological bundles in [26, II].

VI. E_∞ ring spaces and bipermutative categories

An E_∞ space is, essentially, an H-space which is commutative, associative, and unital up to all possible higher coherence homotopies. An E_∞ ring space is, essentially, an E_∞ space with respect to two products, one additive and the other multiplicative, such that the distributive laws are satisfied up to all possible higher coherence homotopies. The precise definition will be given in section 1. Some consequences of the definition, and elementary examples, will be given in section 2.

A symmetric monoidal category is a category with a product which is commutative, associative, and unital up to coherent natural isomorphism. It determines an equivalent permutative category, the classifying space of which is an E_∞ space. A symmetric bimonoidal category is a symmetric monoidal category with respect to two products, one additive and the other multiplicative, such that the distributive laws are satisfied up to coherent natural isomorphism. It determines an equivalent bipermutative category, the classifying space of which is an E_∞ ring space. The precise definitions, and proofs, will be given in sections 3 and 4, and many examples of bipermutative categories will be displayed in section 5.

The relationship between E_∞ ring spaces and E_∞ ring spectra will be determined in chapter VII and applications will be given in chapter VIII. The homology of E_∞ ring spaces has been studied in [26, II].

§1. The definition of E_∞ ring spaces

As will be made precise below, an operad $\mathcal{C}$ is a collection of suitably interrelated spaces $\mathcal{C}(j)$ with actions by the symmetric group Σ_j.

$\mathcal{C}$ is an E_∞ operad if the Σ_j actions are free and the spaces $\mathcal{C}(j)$ are contractible, so that the orbit spaces $\mathcal{C}(j)/\Sigma_j$ are $K(\Sigma_j, 1)$'s. An action θ of $\mathcal{C}$ on a space X in $\mathcal{J}$ (our ground category of based spaces) is a morphism of operads $\mathcal{C} \to \mathcal{E}_X$, where $\mathcal{E}_X(j)$ is the function space $F(X^j, X)$ of based maps and $\mathcal{E}_X$ is given an operad structure in the evident way [45,1.2]. An E_∞ space (X, θ) is a space X together with a given action θ by some E_∞ operad $\mathcal{C}$. With product given by $\theta_2(c): X^2 \to X$ for any $c \in \mathcal{C}(2)$, X is indeed an H-space which is commutative, associative, and unital up to all possible higher coherence homotopies [45, p. 4].

An E_∞ ring space will be an E_∞ space with respect to actions by two interrelated E_∞ operads. These actions will satisfy the distributive laws up to all possible higher coherence homotopies, although these implied homotopies fortunately need not be made explicit.

Thus assume given two operads $\mathcal{C}$ and $\mathcal{G}$. Actions by $\mathcal{C}$ will be thought of as additive and will always be denoted by θ. The corresponding basepoint will be denoted by zero and wedges and smash products will be taken with respect to this basepoint. Actions by $\mathcal{G}$ will be thought of as multiplicative and will always be denoted by ξ. The corresponding basepoint will be denoted by one. As in IV §1, it is convenient to insist that $\mathcal{G}$-spaces have zeros. Recall that $\mathcal{J}_e$ denotes the category of spaces X together with cofibrations $e: S^0 \to X$ and that $\mathcal{G}[\mathcal{J}_e]$ denotes the category of $\mathcal{G}_0$-spaces (X, ξ). It is important to observe that, for non-triviality, zero and one must lie in different components of X. Indeed, in the contrary case, the left translations $\xi_2(g)(0, x)$ and $\xi_2(g)(1, x)$ by zero and one would be homotopic (for any fixed $g \in \mathcal{G}(2)$). Since the first map is trivial and, if $\mathcal{G}(1)$ is connected (as is always the case in practice), the second map is homotopic to the identity [45, p. 4], it follows that X would be contractible.

As explained in [45, §2], an operad $\mathcal{C}$ determines a monad (C, μ, η) in $\mathcal{J}$ such that the notion of a $\mathcal{C}$-space is equivalent to that of a C-algebra in $\mathcal{J}$. We shall define a notion of an action of an operad $\mathcal{G}$ on an operad $\mathcal{C}$ in such a way that C restricts to a monad in the category $\mathcal{G}[\mathcal{J}_e]$. Thus for a $\mathcal{G}_0$-space X, the space CX will inherit a structure of $\mathcal{G}_0$-space such that the unit $\eta: X \to CX$ and product $\mu: CCX \to CX$ will be morphisms of $\mathcal{G}_0$-spaces. We shall then define a $(\mathcal{C}, \mathcal{G})$-space to be an algebra over the monad C in $\mathcal{G}[\mathcal{J}_e]$. The requirement that the additive action $\theta: CX \to X$ be a morphism of $\mathcal{G}_0$-spaces will succinctly encode the distributive laws. An E_∞ ring space will be a $(\mathcal{C}, \mathcal{G})$-space where $\mathcal{C}$ and $\mathcal{G}$ are E_∞ operads.

It is useful to think of the passage from $\mathcal{C}$-spaces to $(\mathcal{C}, \mathcal{G})$-spaces as resulting from a change of ground category from $\mathcal{J}$ to $\mathcal{G}[\mathcal{J}_e]$.

The details of the definitions are necessary for rigor and useful in the study of homology operations [26] and homotopy operations (work in progress). For our present theoretical purposes, it is the conceptual outline above that is crucial. We first recall the definitions of operads and actions by operads.

Notations 1.1. Let $j_r \geq 0$, $1 \leq r \leq k$, and let $j = j_1 + \ldots + j_k$. For $\sigma \in \Sigma_k$, define $\sigma(j_1, \ldots, j_k) \in \Sigma_j$ to be that permutation of j letters which permutes the k blocks of letters determined by the given partition of j as σ permutes k letters. For $\tau_r \in \Sigma_{j_r}$, define $\tau_1 \oplus \ldots \oplus \tau_k \in \Sigma_j$ to be the image of $(\tau_1, \ldots, \tau_k)$ under the evident inclusion of $\Sigma_{j_1} \times \ldots \times \Sigma_{j_k}$ in Σ_j.

Definition 1.2. An operad $\mathcal{C}$ is a collection of spaces $\mathcal{C}(j)$ for $j \geq 0$, with $\mathcal{C}(0)$ a single point $*$, together with maps

$$\gamma: \mathcal{C}(k) \times \mathcal{C}(j_1) \times \ldots \times \mathcal{C}(j_k) \to \mathcal{C}(j_1 + \ldots + j_k)$$

for all $k \geq 0$ and $j_r \geq 0$, a unit element $1 \in \mathcal{C}(1)$, and right actions of Σ_j on $\mathcal{C}(j)$, all subject to the following formulas.

(a) If $c \in \mathcal{C}(k)$, $d_r \in \mathcal{C}(j_r)$ for $1 \leq r \leq k$, and $e_s \in \mathcal{C}(i_s)$ for $1 \leq s \leq j_1 + \ldots + j_k$, then

$$\gamma(\gamma(c; d_1, \ldots, d_k); e_1, \ldots, e_{j_1+\ldots+j_k}) = \gamma(c; f_1, \ldots, f_k),$$

where $f_r = \gamma(d_r; e_{j_1+\ldots+j_{r-1}+1}, \ldots, e_{j_1+\ldots+j_r})$ (or $*$ if $j_r = 0$).

(b) If $c \in \mathcal{C}(k)$, then $\gamma(c; 1^k) = c$; if $d \in \mathcal{C}(j)$, then $\gamma(1; d) = d$.

(c) If $c \in \mathcal{C}(k)$, $d_r \in \mathcal{C}(j_r)$, $\sigma \in \Sigma_k$, and $\tau_r \in \Sigma_{j_r}$, then

$$\gamma(c\sigma; d_1, \ldots, d_k) = \gamma(c; d_{\sigma^{-1}(1)}, \ldots, d_{\sigma^{-1}(k)})\sigma(j_1, \ldots, j_k)$$

and

$$\gamma(c; d_1\tau_1, \ldots, d_k\tau_k) = \gamma(c; d_1, \ldots, d_k)(\tau_1 \oplus \ldots \oplus \tau_k).$$

<u>Definition 1.3.</u> An action θ of an operad $\mathcal{C}$ on a space X consists of Σ_j-equivariant maps $\theta_j: \mathcal{C}(j) \times X^j \to X$ such that $\theta_0(*)$ is the basepoint $* \in X$, $\theta_1(1; x) = x$, and, if $c \in \mathcal{C}(k)$, $d_r \in \mathcal{C}(j_r)$ for $1 \leq r \leq k$, and $x_s \in X$ for $1 \leq s \leq j_1 + \ldots + j_k = j$,

$$\theta_j(\gamma(c; d_1, \ldots, d_k); x_1, \ldots, x_j) = \theta_k(c; y_1, \ldots, y_k),$$

where

$$y_r = \theta_{j_r}(d_r; x_{j_1+\ldots+j_{r-1}+1}, \ldots, x_{j_1+\ldots+j_r}) \quad \text{(or $*$ if } j_r = 0).$$

We require the multiplicative analog of Notations 1.1, and distributivity permuations, to define actions of operads on operads and of operad pairs on spaces.

<u>Notations 1.4.</u> For $j_r \geq 1$, let $S(j_1, \ldots, j_k)$ denote the set of all sequences $I = \{i_1, \ldots, i_k\}$ such that $1 \leq i_r \leq j_r$, and order $S(j_1, \ldots, j_k)$ lexicographi-

cally. This fixes an action of Σ_j on $S(j_1,\ldots,j_k)$, where $j = j_1\cdots j_k$. For $\sigma \in \Sigma_k$, define

$$\sigma<j_1,\ldots,j_k>: S(j_1,\ldots,j_k) \to S(j_{\sigma^{-1}(1)},\ldots,j_{\sigma^{-1}(k)})$$

by

$$\sigma<j_1,\ldots,j_k>\{i_1,\ldots,i_k\} = \{i_{\sigma^{-1}(1)},\ldots,i_{\sigma^{-1}(k)}\}.$$

Via the given isomorphisms of $S(j_1,\ldots,j_k)$ and $S(j_{\sigma^{-1}(1)},\ldots,j_{\sigma^{-1}(k)})$ with $\{1,2,\ldots,j\}$, $\sigma<j_1,\ldots,j_k>$ may be regarded as an element of Σ_j. For $\tau \in \Sigma_{j_r}$, define $\tau_1 \otimes \ldots \otimes \tau_k \in \Sigma_j$ by

$$(\tau_1 \otimes \ldots \otimes \tau_k)\{i_1,\ldots,i_k\} = \{\tau_1 i_1,\ldots,\tau_k i_k\}.$$

<u>Notations 1.5.</u> Given non-negative integers k, j_r for $1 \le r \le k$, and h_{ri} for $1 \le r \le k$ and $1 \le i \le j_r$, define $\nu = \nu\{k, j_r, h_{ri}\}$ to be that permutation of the set of

$$\sum_{I \in S(j_1,\ldots,j_k)} \left(\prod_{r=1}^{k} h_{ri_r}\right) = \prod_{r=1}^{k}\left(\sum_{i=1}^{j_r} h_{ri}\right)$$

letters which corresponds to the comparison of the two ordered sets (where $\coprod$ denotes the ordered disjoint union)

$$\coprod_{I \in S(j_1,\ldots,j_k)} S(h_{1i_1},\ldots,h_{ki_k}) \quad \text{and} \quad S\left(\sum_{i=1}^{j_1} h_{1i},\ldots,\sum_{i=1}^{j_k} h_{ki}\right)$$

obtained by sending an element $\{a_1,\ldots,a_k\}$ with $1 \le a_r \le h_{ri_r}$ of the Ith summand of the first set to that element $\{b_1,\ldots,b_k\}$ of the second set such that $b_r = h_{r1} + \ldots + h_{ri_{r-1}} + a_r$.

<u>Definition 1.6.</u> An action λ of an operad $\mathcal{G}$ on an operad $\mathcal{C}$ consists of maps

$$\lambda: \mathcal{G}(k) \times \mathcal{C}(j_1) \times \ldots \times \mathcal{C}(j_k) \to \mathcal{C}(j_1\cdots j_k)$$

for all $k \ge 0$ and $j_r \ge 0$ subject to the following formulas:

(a) If $g \in \mathcal{G}(k)$, $g_r \in \mathcal{G}(j_r)$ for $1 \leq r \leq k$, and $c_s \in \mathcal{C}(i_s)$ for $1 \leq s \leq j_1 + \ldots + j_k$, then

$$\lambda(\gamma(g; g_1, \ldots, g_k); c_1, \ldots, c_{j_1+\ldots+j_k}) = \lambda(g; d_1, \ldots, d_k),$$

where

$$d_r = \lambda(g_r; c_{j_1+\ldots+j_{r-1}+1}, \ldots, c_{j_1+\ldots+j_r}) \quad (\text{or } * \text{ if } j_r = 0).$$

(a') If $g \in \mathcal{G}(k)$, $c_r \in \mathcal{C}(j_r)$ for $1 \leq r \leq k$, and $d_{ri} \in \mathcal{C}(h_{ri})$ for $1 \leq i \leq j_r$, then

$$\gamma(\lambda(g; c_1, \ldots, c_k); \underset{I \in S(j_1, \ldots, j_k)}{\times} d_I)\nu = \lambda(g; e_1, \ldots, e_k),$$

where

$$d_I = (d_{1i_1}, \ldots, d_{ki_k}) \quad \text{and} \quad e_r = \gamma(c_r; d_{r1}, \ldots, d_{rj_r}).$$

(b) If $c \in \mathcal{C}(j)$ and $1 \in \mathcal{G}(1)$ is the unit of $\mathcal{G}$, then $\lambda(1; c) = c$.

(b') If $g \in \mathcal{G}(k)$ and $1 \in \mathcal{C}(1)$ is the unit of $\mathcal{C}$, then $\lambda(g; 1^k) = 1$.

(c) If $g \in \mathcal{G}(k)$, $c_r \in \mathcal{C}(j_r)$, and $\sigma \in \Sigma_k$, then

$$\lambda(g\sigma; c_1, \ldots, c_k) = \lambda(g; c_{\sigma^{-1}(1)}, \ldots, c_{\sigma^{-1}(k)})\sigma\langle j_1, \ldots, j_k\rangle.$$

(c') If $g \in \mathcal{G}(k)$, $c_r \in \mathcal{C}(j_r)$, and $\tau_r \in \Sigma_{j_r}$, then

$$\lambda(g; c_1\tau_1, \ldots, c_k\tau_k) = \lambda(g; c_1, \ldots, c_k)(\tau_1 \otimes \ldots \otimes \tau_k).$$

Formulas (a), (b), and (c) relate the λ to the internal structure of $\mathcal{G}$ and formulas (a'), (b'), and (c') relate the λ to the internal structure of $\mathcal{C}$.

Definition 1.7. By an operad pair $(\mathcal{C}, \mathcal{G})$, we understand operads $\mathcal{C}$ and $\mathcal{G}$ together with a given action of $\mathcal{G}$ on $\mathcal{C}$. $(\mathcal{C}, \mathcal{G})$ is said to be an E_∞ pair if $\mathcal{C}$ and $\mathcal{G}$ are E_∞ operads. A morphism $(\mathcal{C}, \mathcal{G}) \to (\mathcal{C}', \mathcal{G}')$ of operad pairs is a pair of morphisms of operads $\mathcal{C} \to \mathcal{C}'$ and $\mathcal{G} \to \mathcal{G}'$ which commute with the given actions.

Henceforward, assume given an operad pair $(\mathcal{C}, \mathcal{G})$.

Definition 1.8. Let $(X,\xi) \in \mathcal{G}[\mathcal{J}_e]$. For $k \geq 0$ and $j_r \geq 0$ define maps

$$\xi_k: \mathcal{G}(k) \times \mathcal{C}(j_1) \times X^{j_1} \times \ldots \times \mathcal{C}(j_k) \times X^{j_k} \to \mathcal{C}(j_1 \cdots j_k) \times X^{j_1 \cdots j_k}$$

by the following formula for $g \in \mathcal{G}(k)$, $c_r \in \mathcal{C}(j_r)$, and $y_r \in X^{j_r}$:

$$\xi_k(g, c_1, y_1, \ldots, c_k, y_k) = (\lambda(g; c_1, \ldots, c_k); \bigtimes_{I \in S(j_1, \ldots, j_k)} \xi_k(g; y_I)),$$

where, if $y_r = (x_{r1}, \ldots, x_{rj_r})$, then $y_I = (x_{1i_1}, \ldots, x_{ki_k}) \in X^k$.

Recall the construction of the monad C in $\mathcal{J}$ from [45, 2.4].

Proposition 1.9. For $(X,\xi) \in \mathcal{G}[\mathcal{J}_e]$, the maps ξ_k induce an action of $\mathcal{G}$ on CX such that $\mu: CCX \to CX$ and $\eta: X \to CX$ are morphisms of $\mathcal{G}_0$-spaces. Therefore the monad C in $\mathcal{J}$ restricts to a monad in the category $\mathcal{G}[\mathcal{J}_e]$.

Proof. By (c') and (a') (applied to the degeneracies σ_i of [45, 2.3]), the ξ_k respect the equivalence relation used to define CX. The resulting maps $\xi_k: \mathcal{G}(k) \times (CX)^k \to CX$ satisfy the associativity, unit, and equivariance conditions required of an action by (a), (b), and (c). The map $e: S^0 \to CX$ is the composite of $e: S^0 \to X$ and $\eta: X \to CX$. Now $CCX \in \mathcal{G}[\mathcal{J}_e]$ by iteration, and the maps μ and η commute with the actions ξ by (a') and (b').

Definition 1.10. A $(\mathcal{C}, \mathcal{G})$-space (X, θ, ξ) is a $\mathcal{C}$-space (X, θ) and a $\mathcal{G}_0$-space (X, ξ) such that $\theta: CX \to X$ is a morphism of $\mathcal{G}_0$-spaces. Equivalently, it is required that the diagrams

$$\begin{array}{ccc} \mathcal{G}(k) \times \mathcal{C}(j_1) \times X^{j_1} \times \ldots \times \mathcal{C}(j_k) \times X^{j_k} & \xrightarrow{1 \times \theta_{j_1} \times \ldots \times \theta_{j_k}} & \mathcal{G}(k) \times X^k \\ \downarrow \xi_k & & \downarrow \xi_k \\ \mathcal{C}(j_1 \cdots j_k) \times X^{j_1 \cdots j_k} & \xrightarrow{\theta_{j_1 \cdots j_k}} & X \end{array}$$

commute. A map of $(\mathcal{C}, \mathcal{G})$-spaces is a map which is both a map of $\mathcal{C}$-spaces and a map of $\mathcal{G}_0$-spaces.

Since the notions of $\mathcal{C}$-space and of C-algebra in $\mathcal{J}$ are equivalent [45, 2.8], it follows immediately that the notions of $(\mathcal{C}, \mathcal{G})$-space and of C-algebra in $\mathcal{G}[\mathcal{J}_0]$ are equivalent.

Definition 1.11. An E_∞ ring space is a $(\mathcal{C}, \mathcal{G})$-space where $(\mathcal{C}, \mathcal{G})$ is some E_∞ operad pair.

We have not defined and do not need any notion of a morphism between E_∞ ring spaces over different E_∞ operad pairs.

2. Units; examples of operad pairs

We here point out a few consequences of the definitions and display some examples of actions by one operad on another; these should help to motivate our general theory.

For a space X, CX is the free $\mathcal{C}$-space generated by X. For the same categorical reason [45, 2.9], if X is a $\mathcal{G}_0$-space, then CX is the free $(\mathcal{C}, \mathcal{G})$-space generated by X. In other words, if $(\mathcal{C}, \mathcal{G})[\mathcal{J}_e]$ denotes the category of $(\mathcal{C}, \mathcal{G})$-spaces, then the function

$$\mathrm{Hom}_{\mathcal{G}[\mathcal{J}_e]}((X,\xi),(Y,\xi)) \to \mathrm{Hom}_{(\mathcal{C},\mathcal{G})[\mathcal{J}_e]}((CX,\mu,\xi),(Y,\theta,\xi))$$

obtained by sending $f: X \to Y$ to the composite

$$CX \xrightarrow{Cf} CY \xrightarrow{\theta} Y$$

is a natural isomorphism with inverse $g \to g \circ \eta$ for $g: CX \to Y$.

Definition 2.1. As observed in IV.1.5, (S^0, ξ) is a $\mathcal{G}_0$-space with

$$\xi_j(g)(\varepsilon_1, \ldots, \varepsilon_j) = \varepsilon_1 \cdots \varepsilon_j$$

for $g \in \mathcal{H}(j)$ and $\varepsilon_i = 0$ or 1 (where the product is the ordinary one), and $e: S^0 \to X$ is a morphism of $\mathcal{H}_0$-spaces for any $\mathcal{H}_0$-space X. For a $(\mathcal{C}, \mathcal{H})$-space X, let e also denote the induced map

$$(CS^0, \mu, \xi) \to (X, \theta, \xi)$$

of $(\mathcal{C}, \mathcal{H})$-spaces; e is called the unit of X.

Of course, e is analogous to the unit $Z \to R$ of a ring R.

If G denotes the monad associated to $\mathcal{H}$, then GX is the free $\mathcal{H}$-space generated by a space X. When $\mathcal{C}$ and $\mathcal{H}$ are E_∞ operads, the homology isomorphisms $CX \leftarrow (C \times G)(X) \to GX$ of [45, 3.10 and 46, A.2] show that C and G can be used interchangeably and can thus both be thought of as additive. The $\mathcal{H}$-action on a $(\mathcal{C}, \mathcal{H})$-space gives rise to curious and useful exponential units.

<u>Definition 2.2.</u> Let (X, ξ) be a $\mathcal{H}$-space. For any given element $r \in \pi_0 X$, define a map $e_r: S^0 \to X$ by sending 0 to 1 and 1 to any chosen point in the component r. With S^0 regarded as a based space with base-point 0, form GS^0 and let e_r also denote the induced map

$$(GS^0, \mu) \to (X, \xi)$$

of $\mathcal{H}$-spaces; e_r is called an exponential unit of X. Up to homotopy through $\mathcal{H}$-maps, e_r is independent of the choice of the point in the component r.

Think of the set Z^+ of non-negative integers as the free monoid (under addition) generated by the element 1. The maps e_r for a $(\mathcal{C}, \mathcal{H})$-space are analogous to the maps of monoids from Z^+ into the underlying multiplicative monoid of a ring R.

There are several algebraic examples of operad pairs, for which the component spaces $\mathcal{C}(j)$ and $\mathcal{H}(j)$ are discrete. Let $\mathcal{M}$ and $\mathcal{N}$ denote the operads of [45, 3.1] such that an $\mathcal{M}$-space is a topological monoid and an $\mathcal{N}$-space is a commutative topological monoid. Recall that $\mathcal{M}(j) = \Sigma_j$ with identity element e_j and that $\mathcal{N}(j) = \{f_j\}$.

Lemma 2.3. $(\mathcal{M}, \mathcal{M})$ is an operad pair with respect to the maps $\lambda: \Sigma_k \times \Sigma_j \times \ldots \times \Sigma_{j_k} \to \Sigma_{j_1 \cdots j_k}$ determined by

$$\lambda(e_k; e_{j_1}, \ldots, e_{j_k}) = e_{j_1 \cdots j_k}$$

and the equivariance formulas (c) and (c') of Definition 1.3.

Indeed, the equivariance formulas are dictated by this lemma, which asserts that these formulas are compatible with the distributivity and unit formulas of Definition 1.6. An $(\mathcal{M}, \mathcal{M})$-space is a topological "pseudo semi-ring", "pseudo" meaning that the addition need not be commutative and "semi" meaning that there need not be additive inverses.

Lemma 2.4. For any operad $\mathcal{G}$, $(\mathcal{N}, \mathcal{G})$ is an operad pair with respect to the only possible maps λ, namely

$$\lambda(g; f_{j_1}, \ldots, f_{j_k}) = f_{j_1 \cdots j_k} \quad \text{for all } g \in \mathcal{G}(k).$$

An $(\mathcal{N}, \mathcal{M})$-space is a topological semi-ring and an $(\mathcal{N}, \mathcal{N})$-space is a commutative topological semi-ring. The unique maps $\mathcal{C} \to \mathcal{N}$ and $\mathcal{G} \to \mathcal{N}$ define a morphism $(\mathcal{C}, \mathcal{G}) \to (\mathcal{N}, \mathcal{N})$ of operad pairs for any operad pair $(\mathcal{C}, \mathcal{G})$. Thus any commutative topological semi-ring R is a $(\mathcal{C}, \mathcal{G})$-space. As soon as we verify that E_∞ operad pairs exist, it will follow that any such R is an E_∞ ring space. Conversely, the set $\pi_0 X$ of components of an E_∞ ring space X will be a commutative semi-ring and the discretization map $X \to \pi_0 X$ will be a map of E_∞ ring spaces. Evidently, E_∞ semi-ring space would be a technically more accurate term than E_∞ ring space.

Recall that, algebraically, a commutative semi-ring can be completed to a commutative ring by formation of the Grothendieck group with respect to addition, the multiplication being carried along automatically. Analogously, we shall see in chapter VII that an E_∞ ring space can be completed to an E_∞ ring space which is grouplike with respect to addition, the multiplicative E_∞ structure being carried along automatically. Indeed, this topological completion will induce the algebraic completion on π_0.

The definition of an E_∞ ring space implies the existence of considerable structure on the higher homotopy groups of such a space. The following lemma displays the most obvious bit of structure.

Lemma 2.5. Let (X, θ, ξ) be a $(\mathcal{C}, \mathcal{G})$-space, where $\mathcal{C}$ and $\mathcal{G}$ are locally connected operads. Then the homotopy groups $\pi_* X$ defined with respect to the basepoint zero form a commutative graded semi-ring; if $\pi_0 X$ is a group under addition, then $\pi_* X$ is a ring.

Proof. Addition in $\pi_* X$ is induced by the additive product $\theta_2(c)$ for any $c \in \mathcal{C}(2)$. Since the multiplicative product $\xi_2(g)$ for any $g \in \mathcal{G}(2)$ factors through the smash product $X \wedge X$, it induces pairings $\pi_i X \otimes \pi_j X \to \pi_{i+j} X$ for all $i, j \geq 0$ by letting the product of $\alpha: S^i \to X$ and $\beta: S^j \to X$ be the composite

$$S^{i+j} = S^i \wedge S^j \xrightarrow{\alpha \wedge \beta} X \wedge X \xrightarrow{\xi_2(g)} X.$$

The axioms for a commutative semi-ring follow directly from Definitions 1.6 and 1.8 and the assumption that the $\mathcal{C}(j)$ and $\mathcal{G}(j)$ are connected (compare [45, p. 4]).

The simplest E_∞ operad pair, and the one suited to categorical applications, is derived from $(\mathcal{M}, \mathcal{N})$ by application of the product-preserving functor $|D_*(?)|$ from spaces to contractible spaces given by [45, 10.2 and 11.1]. Recall from [45, 15.1 and 10.3] that application of this functor to $\mathcal{M}$ gives an E_∞ operad $\mathcal{Q}$ such that $\mathcal{Q}(j)$ is just the normalized version of Milnor's universal Σ_j-bundle.

<u>Lemma 2.6.</u> $(\mathcal{Q}, \mathcal{Q})$ is an E_∞ operad pair; the action λ of $\mathcal{Q}$ on itself is obtained by application of the functor $|D_*(?)|$ to the action of $\mathcal{M}$ on itself.

<u>Proof.</u> The formulas of Definition 1.6 can be written out as commutative diagrams, hence, by functoriality, these formulas hold for $\mathcal{Q}$ since they do so for $\mathcal{M}$.

A categorical description of $(\mathcal{Q}, \mathcal{Q})$ will be given in section 4. The following remarks will be needed in VII § 3.

<u>Remarks 2.7.</u> (i) $\Sigma_j = D_0(\Sigma_j)$, and there results an inclusion of operads $\mathcal{M} \subset \mathcal{Q}$ and an inclusion of operad pairs $(\mathcal{M}, \mathcal{M}) \subset (\mathcal{Q}, \mathcal{Q})$. Thus a $\mathcal{Q}$-space is also a topological monoid (with product $\oplus$) and a $(\mathcal{Q}, \mathcal{Q})$-space is also a topological pseudo semi-ring (with second product $\otimes$). The products $\oplus$ and $\otimes$ coincide with those given in terms of the actions as $\theta_2(e_2)$ and $\xi_2(e_2)$.

(ii) Ligaard and Madsen [39, 2.2] have verified that any $\mathcal{Q}$-space Y is a strongly homotopy commutative H-space with respect to the product $\oplus$. Therefore the classifying space BY is an H-space and the natural map $\zeta : Y \to \Omega BY$ is a group completion in the sense of [46, 2.1] (e.g., by [47, 15.1]).

(iii) It is sometimes convenient to replace general E_∞ spaces by equivalent $\mathcal{Q}$-spaces. This can be done as follows. Given the $\mathcal{C}$-space (X, θ), where $\mathcal{C}$ is an E_∞ operad, construct the maps

$$X \xleftarrow{\varepsilon(\theta\pi_1)} B(C\times D, C\times D, X) \xrightarrow{B(\pi_2, 1, 1)} B(D, C\times D, X).$$

Here the bar constructions are specified by [45, 9.6 and 11.1 (see p. 126)], $B(D, C \times D, X)$ is a $\mathcal{Q}$-space and both maps are morphisms of $\mathcal{C} \times \mathcal{Q}$-spaces by [45, 12.2], $\varepsilon(\theta\pi_1)$ is a strong deformation retraction with right inverse $\tau(\eta)$ by [45, 9.8 and 11.10], and $B(\pi_2, 1, 1)$ is a homotopy equivalence by [46, A.2 (ii) and A.4(ii)].

(iv) For X as in (iii), let $GX = \Omega BB(D, C \times D, X)$ and let $g = \zeta \circ B(\pi_2, 1, 1) \circ \tau(\eta): X \to GX$. By (ii), g is a natural group completion of the $\mathcal{C}$-space X. The existence of such a construction was asserted in [46, 2.1], but the argument given there was incomplete.

(v) In the second result labelled Theorem 3.7 in [46], I claimed that $\oplus$ on DX was a morphism of $\mathcal{Q}$-spaces. That assertion is clearly false, as it would imply that $\oplus$ is actually commutative. The mistake occurs in the formula for γ [46, p. 76], from which a factor $\sigma(j_1, \ldots, j_k)$ was omitted (compare section 4).

(vi) Aside from use of (iv) in the proof of VII. 3.1, we shall ignore the classifying spaces which result from the monoid structures on $\mathcal{Q}$-spaces and $(\mathcal{Q}, \mathcal{Q})$-spaces in favor of the deloopings constructed by application of the machinery of chapter VII. The latter have numerous special properties essential to our theory, and I have not proven that the two are equivalent.

§3. Symmetric bimonoidal and bipermutative categories

Categories with appropriate internal structure provide a very rich source of E_∞ spaces and E_∞ ring spaces. Here all categories with internal structure are to be small and topological, and all functors and natural transformations are to be continuous. For a category $\mathcal{A}$,

$\mathcal{O}\mathcal{A}$ and $\mathcal{M}\mathcal{A}$ denote the spaces of objects and morphisms of $\mathcal{A}$ and S, T, I, and C denote the source, target, identity, and composition functions, all of which are required to be continuous. If no topology is in sight, we can always impose the discrete topology.

Recall that a symmetric monoidal category is a category $\mathcal{A}$ together with a functor $\square : \mathcal{A} \times \mathcal{A} \to \mathcal{A}$ and an object $*$ such that $\square$ is associative, (right) unital, and commutative up to coherent natural isomorphisms a, b, and c [40, VII, §1 and §7]. $\mathcal{A}$ is permutative if $\square$ is strictly associative and unital, with no isomorphisms required. Coherence with the remaining piece of structure, the commutativity isomorphism c, is then guaranteed by commutativity of the following diagrams for $A, B, C \in \mathcal{O}\mathcal{A}$:

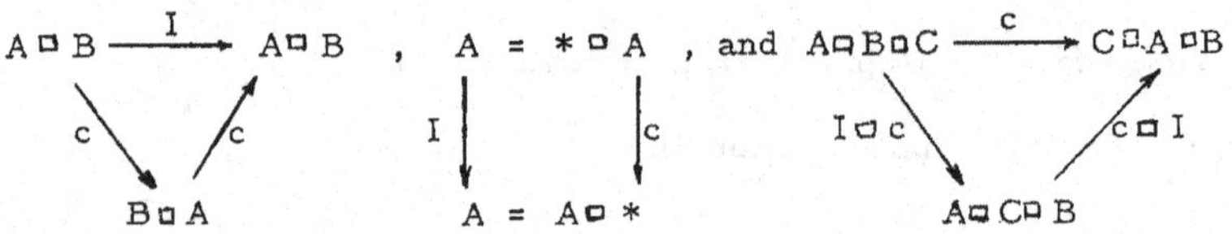

$$\begin{array}{ccc} A \square B & \xrightarrow{I} & A \square B \\ & {\scriptstyle c}\searrow \quad \nearrow{\scriptstyle c} & \\ & B \square A & \end{array} \quad , \quad \begin{array}{ccc} A & = & * \square A \\ {\scriptstyle I}\downarrow & & \downarrow{\scriptstyle c} \\ A & = & A \square * \end{array} \quad , \text{ and } \quad \begin{array}{ccc} A \square B \square C & \xrightarrow{c} & C \square A \square B \\ & {\scriptstyle I \square c}\searrow \quad \nearrow{\scriptstyle c \square I} & \\ & A \square C \square B & \end{array}$$

Symmetric monoidal categories can be replaced functorially by naturally equivalent permutative categories, but the relevant notions of morphism require explanation. This is particularly so since the usual categorical definition of a coherent functor between symmetric monoidal categories would allow examples like the forgetful functor from modules over a commutative ring R under $\otimes_R$ to Abelian groups under $\otimes_Z$ and is too lax for our purposes.

<u>Definition 3.1.</u> A morphism $\mathcal{A} \to \mathcal{A}'$ of symmetric monoidal categories is a functor $F: \mathcal{A} \to \mathcal{A}'$ such that $F* = *$ together with a natural isomorphism $\phi: FA \square FB \to F(A \square B)$ such that the following diagrams are commutative:

$$\begin{array}{ccccc}
(FA \,\square\, FB) \,\square\, FC & \xrightarrow{a} & FA \,\square\, (FB \,\square\, FC) & \xrightarrow{1 \square \phi} & FA \,\square\, F(B \,\square\, C) \\
\downarrow{\scriptstyle \phi \square 1} & & & & \downarrow{\scriptstyle \phi} \\
F(A \,\square\, B) \,\square\, FC & \xrightarrow{\phi} & F((A \,\square\, B) \,\square\, C) & \xrightarrow{Fa} & F(A \,\square\, (B \,\square\, C))
\end{array}$$

$$\begin{array}{ccc}
FA & \xrightarrow{b} & FA \,\square\, F* \\
\downarrow{\scriptstyle I} & & \downarrow{\scriptstyle \phi} \\
FA & \xrightarrow{Fb} & F(A \,\square\, *)
\end{array}
\quad \text{and} \quad
\begin{array}{ccc}
FA \,\square\, FB & \xrightarrow{c} & FB \,\square\, FA \\
\downarrow{\scriptstyle \phi} & & \downarrow{\scriptstyle \phi} \\
F(A \,\square\, B) & \xrightarrow{Fc} & F(B \,\square\, A)
\end{array}$$

A morphism $\mathcal{A} \to \mathcal{A}'$ of permutative categories is a functor $F: \mathcal{A} \to \mathcal{A}'$ such that $F* = *$, $FA \,\square\, FB = F(A \,\square\, B)$, and $c = Fc$ on $FA \,\square\, FB = F(A \,\square\, B)$. Note that a morphism of symmetric monoidal categories between permutative categories need not be a morphism of permutative categories.

A slight elaboration of the proof of [46, 4.2] gives the following more precise result.

<u>Proposition 3.2.</u> There is a functor Φ from the category of symmetric monoidal categories to the category of permutative categories and a natural equivalence $\pi: \Phi\mathcal{A} \to \mathcal{A}$ of symmetric monoidal categories. If $\mathcal{A}$ is permutative, then π is a morphism of permutative categories.

One often encounters categories with two symmetric monoidal structures, one additive and one multiplicative, which satisfy the (right) distributive and nullity of zero laws up to coherent natural isomorphisms d and n. We shall say that such a category is symmetric bimonoidal. Laplaza [35] has made a careful study of such categories. In particular, he has given a list of diagrams the commutativity of which ensures that all further coherence diagrams which can reasonably be expected to commute do in fact commute. Comparison of his list with the notion of an E_∞ ring space leads to the following definition.

Definition 3.3. A bipermutative category $(\mathcal{A}, \oplus, 0, c, \otimes, 1, \tilde{c})$ is a pair of permutative categories $(\mathcal{A}, \oplus, 0, c)$ and $(\mathcal{A}, \otimes, 1, \tilde{c})$ such that the following three conditions are satisfied.

(i) $0 \otimes A = 0 = A \otimes 0$ for $A \in \mathcal{O}\mathcal{A}$ and $I(0) \otimes f = I(0) = f \otimes I(0)$ for $f \in \mathcal{M}\mathcal{A}$; that is, 0 is a strict two-sided zero object for $\otimes$.

(ii) The right distributive law is strictly satisfied by objects and morphisms, and the following diagram commutes for $A, B, C \in \mathcal{O}\mathcal{A}$:

$$\begin{array}{ccc} (A \oplus B) \otimes C & = & (A \otimes C) \oplus (B \otimes C) \\ \Big\downarrow c \otimes 1 & & \Big\downarrow c \\ (B \oplus C) \otimes C & = & (B \otimes C) \oplus (A \otimes C) \end{array} .$$

(iii) Define a natural left distributivity isomorphism ℓ as the following composite

$$A \otimes (B \oplus C) \xrightarrow{\tilde{c}} (B \oplus C) \otimes A = (B \otimes A) \oplus (C \otimes A) \xrightarrow{\tilde{c} \times \tilde{c}} (A \otimes B) \oplus (A \otimes C);$$

then the following diagram commutes for $A, B, C, D \in \mathcal{O}\mathcal{A}$:

$$\begin{array}{ccc} (A \oplus B) \otimes (C \oplus D) & = & (A \otimes (C \oplus D)) \oplus (B \otimes (C \oplus D)) \\ \Big\downarrow \ell & & \Big\downarrow \ell \oplus \ell \\ ((A \oplus B) \otimes C) \oplus ((A \oplus B) \otimes D) & & (A \otimes C) \oplus (A \otimes D) \oplus (B \otimes C) \oplus (B \otimes D) \\ & \searrow\!\!= \quad (A \otimes C) \oplus (B \otimes C) \oplus (A \otimes D) \oplus (B \otimes D) \quad \swarrow 1 \oplus c \oplus 1 & \end{array}$$

Laplaza's work [35, p. 40] implies that a bipermutative category is symmetric bimonoidal. In the absence of strict commutativity, it is clearly unreasonable to demand that both distributive laws hold strictly. The choice of which law to make strict is logically arbitrary, but our choice is dictated by consistency with the lexicographic ordering used in Notations 1.4.

Definition 3.4. A morphism $\mathcal{A} \to \mathcal{A}'$ of symmetric bimonoidal categories is a functor $F: \mathcal{A} \to \mathcal{A}'$ such that $F0 = 0$ and $F1 = 1$ together with natural isomorphisms $\phi: FA \oplus FB \to F(A \oplus B)$ and $\psi: FA \otimes FB \to F(A \otimes B)$ such that (F, ϕ) and (F, ψ) are morphisms of symmetric monoidal categories and the following diagrams are commutative:

$$\begin{array}{ccc} F0 & \xrightarrow{n} & FA \otimes F0 \\ {\scriptstyle I}\downarrow & & \downarrow{\scriptstyle \psi} \\ F0 & \xrightarrow{Fn} & F(A \otimes 0) \end{array} \quad \text{and} \quad \begin{array}{ccc} (FA \oplus FB) \otimes FC & \xrightarrow{d} & (FA \otimes FC) \oplus (FB \otimes FC) \\ {\scriptstyle \phi \otimes 1}\downarrow & & \downarrow{\scriptstyle \psi \oplus \psi} \\ F(A \oplus B) \otimes FC & & F(A \otimes C) \oplus F(B \otimes C) \\ {\scriptstyle \psi}\downarrow & & \downarrow{\scriptstyle \phi} \\ F((A \oplus B) \otimes C) & \xrightarrow{Fd} & F((A \otimes C) \oplus (B \otimes C)) \end{array}$$

A morphism $\mathcal{A} \to \mathcal{A}'$ of bipermutative categories is a functor $F: \mathcal{A} \to \mathcal{A}'$ which is a morphism of permutative categories with respect to both the additive and multiplicative structures. Again, a morphism of symmetric bimonoidal categories between bipermutative categories need not be a morphism of bipermutative categories.

Proposition 3.5. There is a functor Φ from the category of symmetric bimonoidal categories to the category of bipermutative categories and a natural equivalence $\pi: \Phi\mathcal{A} \to \mathcal{A}$ of symmetric bimonoidal categories. If $\mathcal{A}$ is bipermutative, then π is a morphism of bipermutative categories.

Proof. To avoid technical topological difficulties, assume either that $\mathcal{O}\mathcal{A}$ is discrete (which is the case in practice) or that 0 and 1 are non-degenerate basepoints such that 0 is a strict unit for $\oplus$ and 1 is a strict unit for $\otimes$. The latter condition can always be arranged by growing whiskers on the given 0 and 1 (by adjoining copies of the category $\mathcal{J}$ with two objects and one non-identity morphism) so as to obtain a new 0 and 1

as required. We construct $\mathcal{B} = \Phi\mathcal{A}$ as follows. Let $(\mathcal{O}\mathcal{A})'$ be the free topological monoid, with product denoted by $\boxtimes$, generated by $\mathcal{O}\mathcal{A}$ modulo the relations $e = 1$ and $0 \boxtimes A = 0 = A \boxtimes 0$ for all $A \in \mathcal{O}\mathcal{A}$. Let $\mathcal{O}\mathcal{B}$ be the free topological monoid, with product denoted by $\boxplus$, generated by $(\mathcal{O}\mathcal{A})'$ modulo the relation $e = 0$. Extend the product $\boxtimes$ from $(\mathcal{O}\mathcal{A})'$ to all of $\mathcal{O}\mathcal{B}$ by the formula

$$(A_1 \boxplus \dots \boxplus A_m) \boxtimes (B_1 \boxplus \dots \boxplus B_n)$$

$$= (A_1 \boxtimes B_1) \boxplus \dots \boxplus (A_1 \boxtimes B_n) \boxplus \dots \boxplus (A_m \boxtimes B_1) \boxplus \dots \boxplus (A_m \boxtimes B_n)$$

for $A_i, B_j \in (\mathcal{O}\mathcal{A})'$. Both products on $\mathcal{O}\mathcal{B}$ are associative, 0 is a strict unit for $\boxplus$ and zero for $\boxtimes$, 1 is a strict unit for $\boxtimes$, and the right distributive law holds. Let $\eta: \mathcal{O}\mathcal{A} \to \mathcal{O}\mathcal{B}$ denote the evident inclusion. Define $\pi: \mathcal{O}\mathcal{B} \to \mathcal{O}\mathcal{A}$ by $\pi(0) = 0$, $\pi(1) = 1$,

$$\pi(A_1 \boxtimes \dots \boxtimes A_n) = A_1 \otimes (A_2 \otimes (A_3 \otimes \dots (A_{n-1} \otimes A_n) \dots))$$

for $A_i \in \mathcal{O}\mathcal{A}$, $A_i \neq 0$ and $A_i \neq 1$, and

$$\pi(A'_1 \boxplus \dots \boxplus A'_n) = \pi A'_1 \oplus (\pi A'_2 \oplus (\pi A'_3 \oplus \dots (\pi A'_{n-1} \oplus \pi A'_n) \dots))$$

for $A'_i \in (\mathcal{O}\mathcal{A})'$, $A'_i \neq 0$. Define $\mathcal{M}\mathcal{B}$ by

$$\mathcal{B}(B, B') = \{B\} \times \mathcal{A}(\pi B, \pi B') \times \{B'\} .$$

The singleton sets determine S and T for $\mathcal{B}$, and I and C are induced from the corresponding functions for $\mathcal{A}$. $\mathcal{M}\mathcal{B}$ is topologized as a subspace of $\mathcal{O}\mathcal{B} \times \mathcal{M}\mathcal{A} \times \mathcal{O}\mathcal{B}$. The products $\boxplus$ and $\boxtimes$ on $\mathcal{M}\mathcal{B}$ and the symmetries c and $\tilde{c}$ of $\mathcal{B}$ are determined by the following arrows of $\mathcal{A}$:

$$\pi(B \boxplus C) \xrightarrow{\cong} \pi B \oplus \pi C \xrightarrow{f \oplus g} \pi B' \oplus \pi C' \xrightarrow{\cong} \pi(B' \boxplus C')$$

$$\pi(B \boxtimes C) \xrightarrow{\cong} \pi B \otimes \pi C \xrightarrow{f \otimes g} \pi B' \otimes \pi C' \xrightarrow{\cong} \pi(B' \boxtimes C')$$

for morphisms (B, f, B') and (C, g, C') of $\mathcal{B}$ and

$$\pi(B \boxplus C) \xrightarrow{\cong} \pi B \oplus \pi C \xrightarrow{c} \pi C \oplus \pi B \xrightarrow{\cong} \pi(C \boxplus B)$$

$$\pi(B \boxtimes C) \xrightarrow{\cong} \pi B \otimes \pi C \xrightarrow{\tilde{c}} \pi C \otimes \pi B \xrightarrow{\cong} \pi(C \boxtimes B)$$

for objects B and C of $\mathcal{B}$; the unlabelled isomorphisms are uniquely determined by the monoidal structures of $\mathcal{A}$. Define $\eta : \mathcal{M}\mathcal{A} \to \mathcal{M}\mathcal{B}$ by $\eta(f) = (A, f, A')$ for $f: A \to A'$ and define $\pi: \mathcal{M}\mathcal{B} \to \mathcal{M}\mathcal{A}$ by $\pi(B, g, B') = g$ for $g: \pi B \to \pi B'$. Then η and π are functors, $\pi\eta$ is the identity functor, and the morphisms $(B, I\pi B, \eta\pi B)$ of $\mathcal{B}$ define a natural isomorphism between $\eta\pi$ and the identity functor of $\mathcal{B}$. The remaining verifications are equally straightforward.

§4. Bipermutative categories and E_∞ ring spaces

We here describe the E_∞ operad pair $(\mathcal{Q}, \mathcal{Q})$ categorically, review the passage from permutative categories to $\mathcal{Q}$-spaces obtained in [46, §4], and construct a functor from the category of bipermutative categories to the category of $(\mathcal{Q}, \mathcal{Q})$-spaces.

Recall that the translation category $\tilde{G}$ of a monoid G has objects the elements of G and morphisms from g' to g'' those elements $g \in G$ such that $g'g = g''$. When G is a group, g is unique and a functor with range $\tilde{G}$ is therefore uniquely determined by its object function. Note that G acts from the right on $\tilde{G}$ via the product of G.

Let $\tilde{\gamma}: \tilde{\Sigma}_k \times \tilde{\Sigma}_{j_1} \times \dots \times \tilde{\Sigma}_{j_k} \to \tilde{\Sigma}_{j_1+\dots+j_k}$ be the functor defined on objects by the formula

$$(1) \qquad \tilde{\gamma}(\sigma; \tau_1, \dots, \tau_k) = (\tau_{\sigma^{-1}(1)} \oplus \dots \oplus \tau_{\sigma^{-1}(k)})\sigma(j_1, \dots, j_k).$$

(The factor $\sigma(j_1, \dots, j_k)$ was inadvertently omitted from the definition of $\tilde{\gamma}$ given in [46, p. 82].)

Let $\tilde{\lambda}: \tilde{\Sigma}_k \times \tilde{\Sigma}_{j_1} \times \ldots \times \tilde{\Sigma}_{j_k} \to \tilde{\Sigma}_{j_1 \ldots j_k}$ be the functor defined on objects by the formula

$$(2) \qquad \tilde{\lambda}(\sigma; \tau_1, \ldots, \tau_k) = (\tau_{\sigma^{-1}(1)} \otimes \ldots \otimes \tau_{\sigma^{-1}(k)})\sigma\langle j_1, \ldots, j_k\rangle .$$

Let $B\mathcal{A}$ denote the classifying space of a (small, topological) category $\mathcal{A}$ and recall that B is a product-preserving functor from categories to spaces (e.g. [46, 4.6]). As observed in [46, 4.7], $B\tilde{G}$ coincides with $|D_*G|$ for any topological group G. By comparison of (1) to the equivariance formulas in Definition 1.2, the structural maps γ of the E_∞ operad $\mathcal{Q}$ coincide with the maps

$$B\tilde{\gamma}: B\tilde{\Sigma}_k \times B\tilde{\Sigma}_{j_1} \times \ldots \times B\tilde{\Sigma}_{j_k} \to B\tilde{\Sigma}_{j_1 + \ldots + j_k} .$$

By comparison of (2) to the equivariance formulas in Definition 1.6, the maps λ which give the action of $\mathcal{Q}$ on itself coincide with the maps

$$B\tilde{\lambda} : B\tilde{\Sigma}_k \times B\tilde{\Sigma}_{j_1} \times \ldots \times B\tilde{\Sigma}_{j_k} \to B\tilde{\Sigma}_{j_1 \ldots j_k} .$$

Alternatively, this description can be used to define the E_∞ operad pair $(\mathcal{Q}, \mathcal{Q})$.

Let $(\mathcal{A}, \square, *, c)$ be a permutative category. As pointed out in [46, p. 81], c determines Σ_j-equivariant functors

$$c_j : \tilde{\Sigma}_j \times \mathcal{A}^j \to \mathcal{A}$$

such that c_j restricts to the j-fold iterate of $\square$ on $\{e_j\} \times \mathcal{A}^j = \mathcal{A}^j$. The coherence diagrams of the previous section imply the following result. Indeed, by the very meaning of coherence, we need only observe that the diagram of the lemma makes sense on objects.

<u>Lemma 4.1.</u> The following diagram is commutative for all $j \geq 0$, $k \geq 0$, and $j_i \geq 0$ such that $j_1 + \ldots + j_k = j$:

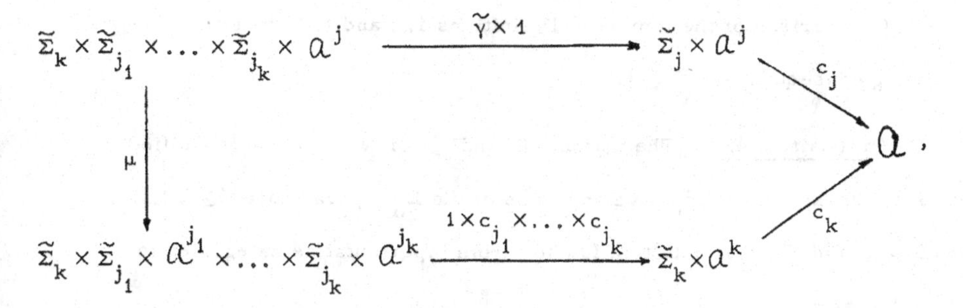

where μ is the evident shuffle isomorphism of categories.

Comparison of the lemma to Definition 1.3 gives the following consequence.

<u>Proposition 4.2.</u> Define $\theta_j = Bc_j : \mathcal{D}(j) \times (B\mathcal{A})^j = B(\tilde{\Sigma}_j \times \mathcal{A}^j) \to B\mathcal{A}$. Then the θ_j define an action θ of $\mathcal{D}$ on $B\mathcal{A}$, and B restricts to a functor from permutative categories to $\mathcal{D}$-spaces.

Now let $(\mathcal{A}, \oplus, 0, c, \otimes, 1, \tilde{c})$ be a bipermutative category. Then coherence implies the following analog of Lemma 4.1.

<u>Lemma 4.3.</u> The following diagram is commutative for all $j \geq 0$, $k \geq 0$, and $j_i \geq 0$ such that $j_1 \cdots j_k = j$:

$$\begin{array}{ccccc}
\tilde{\Sigma}_k \times \tilde{\Sigma}_{j_1} \times \mathcal{A}^{j_1} \times \cdots \times \tilde{\Sigma}_{j_k} \times \mathcal{A}^{j_k} & \xrightarrow{1 \times c_{j_1} \times \cdots \times c_{j_k}} & \tilde{\Sigma}_k \times \mathcal{A}^k & & \\
 & & & \searrow^{\tilde{c}_k} & \\
\downarrow \omega & & & & \mathcal{A}, \\
 & & & \nearrow_{c_j} & \\
\tilde{\Sigma}_k \times \tilde{\Sigma}_{j_1} \times \cdots \times \tilde{\Sigma}_{j_k} \times (\tilde{\Sigma}_k \times \mathcal{A}^k)^j & \xrightarrow{\tilde{\lambda} \times \tilde{c}_k^j} & \tilde{\Sigma}_j \times \mathcal{A}^j & &
\end{array}$$

where ω is defined on objects and morphisms by the formula

$$\omega(\sigma, \tau_1, y_1, \ldots, \tau_k, y_k) = (\sigma, \tau_1, \ldots, \tau_k, \underset{I \in S(j_1, \ldots, j_k)}{\times} (\sigma, y_I)).$$

Comparison of the lemma to Definitions 1.5 and 1.7 gives the following consequence.

Proposition 4.4. The actions θ and ξ of $\mathcal{D}$ on $B\mathcal{A}$ specified by $\theta_j = Bc_j$ and $\xi_j = B\check{c}_j$ give an action of the E_∞ operad pair $(\mathcal{D}, \mathcal{D})$ on $B\mathcal{A}$, and B restricts to a functor from bipermutative categories to $(\mathcal{D}, \mathcal{D})$-spaces.

The following addendum is sometimes useful.

Remarks 4.5. For $i \in \mathcal{O}\mathcal{A}$, let $\mathcal{A}_i$ denote the subcategory of $\mathcal{A}$ which contains the unique object i and all morphisms from i to i; $B\mathcal{A}_i$ is the ordinary classifying space of this monoid of morphisms. Clearly $(B\mathcal{A}_0, \theta)$ is a sub $\mathcal{D}$-space of $(B\mathcal{A}, \theta)$ and $(B\mathcal{A}_1, \xi)$ is a sub $\mathcal{D}$-space of $(B\mathcal{A}, \xi)$.

There is a more general way of looking at the constructions above. One can think of $\{D_*\Sigma_j | j \geq 0\}$ as specifying an operad $\mathcal{D}_*$ in the category of simplicial spaces (or sets, since the $D_q\Sigma_j$ are discrete). The actions of $\mathcal{D}$ on $B\mathcal{A}$ result by passage to geometric realization [45, §11] from actions of $\mathcal{D}_*$ on the simplicial spaces $B_*\mathcal{A}$ [46, 4.6]. Although no such examples will be studied in this volume, there exist simplicial spaces with actions by $\mathcal{D}_*$ or by the pair $(\mathcal{D}_*, \mathcal{D}_*)$ which are not of the form $B_*\mathcal{A}$ for any $\mathcal{A}$; clearly our theory will apply to their realizations.

§5. Examples of bipermutative categories

The seminal example, which will map into all others, is the following one.

Example 5.1. Let $\mathcal{E}$ denote the category of finite sets $n \geq 0$ and their isomorphisms. Think of n as $\{1, 2, \ldots, n\}$ and identify $\mathcal{E}(n, n)$ with the symmetric group Σ_n. Then $(\mathcal{E}, \oplus, 0, c, \otimes, 1, \tilde{c})$ is a bipermutative category, where $\oplus$ and $\otimes$ are defined are defined on objects and morphisms by

$$m \oplus n = m + n \quad \text{and} \quad (\sigma \oplus \tau)(i) = \begin{cases} \sigma(i) & \text{if } 1 \leq i \leq m \\ m + \tau(i-m) & \text{if } m < i \leq m+n \end{cases}$$

and

$$m \otimes n = mn \quad \text{and} \quad (\sigma \otimes \tau)((i-1)n + j) = ((\sigma(i) - 1)n + \tau(j)), \quad 1 \leq i \leq m \text{ and } 1 \leq j \leq n,$$

and where $c = c(m, n) \in \Sigma_{m+n}$ and $\tilde{c} = \tilde{c}(m, n) \in \Sigma_{mn}$ are defined by

$$c(i) = \begin{cases} n+i & \text{if } 1 \leq i \leq m \\ i-m & \text{if } m < i \leq m+n \end{cases}$$

and

$$\tilde{c}((i-1)n+j) = (j-1)m + i, \quad 1 \leq i \leq m \quad \text{and} \quad 1 \leq j \leq n.$$

For $\otimes$, mn should be thought of as $\{(1,1), \ldots, (1,n), \ldots, (m,1), \ldots, (m,n)\}$ and it is this choice of order (required for consistency with Notations 1.4) which leads to the strict right, rather than left, distributive law. The space DS^0 is the disjoint union of the orbit spaces $\mathcal{Q}(j)/\Sigma_j$ [45, 8.11], and the unit $e: DS^0 \to B\mathcal{E}$ of Definition 2.1 coincides with the disjoint union of the homeomorphisms

$$|D_*\Sigma_j|/\Sigma_j \to E\Sigma_j/\Sigma_j = B\Sigma_j$$

specified in [45, 10.3]. (Alternatively, use $E\Sigma_j = B\tilde{\Sigma}_j$.) We therefore regard e as an identification and conclude that $B\mathcal{E}$ is the free $(\mathcal{Q}, \mathcal{Q})$-space generated by S^0.

For a topological ring A, the groups $K_i A$ for $i > 0$ can be defined in terms of permutative categories $\mathcal{P}A$ or $\mathcal{F}A$ of finitely generated projective or free left A-modules (as will be discussed in VIII §1). When A is commutative, $\mathcal{P}A$ and $\mathcal{F}A$ can be taken as bipermutative categories. In the case of $\mathcal{F}A$, we can be more explicit.

Example 5.2. Define $\mathcal{GL}A$ as follows. The objects of $\mathcal{GL}A$ are the non-negative integers, each thought of as A^n together with its standard ordered basis $\{e_1, \ldots, e_n\}$. The morphisms of $\mathcal{GL}A$ are the isomorphisms $A^n \to A^n$. Thus $\mathcal{GL}A(m,n)$ is empty if $m \neq n$ and $\mathcal{GL}A(n,n) = GL(n,A)$. Define a functor $e\colon \mathcal{E} \to \mathcal{GL}A$ by $e(n) = n$ on objects and $e(\sigma)(e_i) = e_{\sigma^{-1}(i)}$ on morphisms $\sigma \in \Sigma_n$. Then $(\mathcal{GL}A, \oplus, 0, c)$ is a permutative category and, if A is commutative, $(\mathcal{GL}A, \oplus, 0, c, \otimes, 1, \tilde{c})$ is a bipermutative category, where $\oplus, \otimes$, c, and $\tilde{c}$ are specified by the requirements that $\oplus$ and $\otimes$ have their usual meanings (with respect to the isomorphisms $A^m \oplus A^n \to A^{m+n}$ and $A^m \otimes A^n \to A^{mn}$ defined as usual on ordered bases) and that $e\colon \mathcal{E} \to \mathcal{GL}A$ be a morphism of permutative and, if A is commutative, bipermutative categories. Note that e factors through $\mathcal{GL}Z$ for any A. By the naturality of the unit of $(\mathcal{Q}, \mathcal{Q})$-spaces, $Be\colon B\mathcal{E} \to B\mathcal{GL}A$ coincides (under the identification $DS^0 = B\mathcal{E}$) with the unit of $B\mathcal{GL}A$. We identify morphisms of $\mathcal{GL}A$ with their matrices with respect to the standard bases. For a morphism of rings $\alpha\colon A \to A'$, define $\mathcal{GL}\alpha\colon \mathcal{GL}A \to \mathcal{GL}A'$ by applying α to all entries of matrices. Then $\mathcal{GL}$ is a functor from rings to permutative categories and from commutative rings to bipermutative categories.

Example 5.3. If A is commutative, define $\mathcal{O}A$ to be the sub bipermutative category of $\mathcal{GL}A$ whose morphisms are the orthogonal matrices $(MM^t = I)$. Then $\mathcal{O}$ is also a functor from commut ative rings to bipermutative categories. Although we cannot simply restrict morphisms to matrices of determinant one, since the permutation matrices required to define c and $\check{c}$ would no longer be present, we can obtain a sub permutative category $\mathcal{SGL}$ of $\mathcal{GL}$ under $\oplus$ by restricting objects to the even numbers and restricting morphisms to the elements of SGL(2n, A). Similarly, define $\mathcal{SO} = \mathcal{O} \cap \mathcal{SGL}$.

In the examples above, the set of $n \times n$-matrices with entries in A is to be given the obvious product topology and GL(n, A) and O(n, A) are to be given the subspace topologies. We have insisted that rings be topologized in order to treat algebraic and topological K-theory simultaneously.

<u>Example 5.4.</u> Let $\mathbb{R}$, $\mathbb{C}$, and $\mathbb{H}$ be the (topologized) real numbers, complex numbers, and quaternions. Define subcategories $\mathcal{O} = \mathcal{O}\mathbb{R}$, $\mathcal{U}$ and $\mathcal{S}p$ of $\mathcal{GL}\mathbb{R}$, $\mathcal{GL}\mathbb{C}$, and $\mathcal{GL}\mathbb{H}$ by restricting to orthogonal, unitary and symplectic linear transformations, respectively. Then $\mathcal{O}$ and $\mathcal{U}$ are bipermutative categories, and complexification $\mathcal{O} \to \mathcal{U}$ is a morphism of bipermutative categories. $\mathcal{S}p$ is an (additive) permutative category and symplectification $\mathcal{U} \to \mathcal{S}p$ is a morphism of permutative categories. When appropriately specified on bases, the forgetful functors $\mathcal{S}p \to \mathcal{U}$ and $\mathcal{U} \to \mathcal{O}$ are morphisms of (additive) permutative categories, with object functions which send n to 2n.

The following three examples, whose significance was first understood by Quillen [58, 59] and Tornehave [75, 77], are central to the interplay between algebraic and topological K-theory to be discussed in chapter VIII.

<u>Example 5.5.</u> For a perfect field k of characteristic $q \neq 0$, let

$$\phi^q : \mathcal{GL}k \to \mathcal{GL}k \quad \text{and} \quad \phi^q : \mathcal{O}k \to \mathcal{O}k$$

denote the morphisms of bipermutative categories derived from the Frobenius automorphism $x \to x^q$ of k. For $r = q^a$, let ϕ^r be the a-fold iterate of ϕ^q. This example is most interesting when k is the algebraic closure of the field of q elements.

Example 5.6. Let $r = q^a$ where q is a prime and $a \geq 1$. Let k_r be a field with r elements. Define a forgetful functor $f: \mathcal{GL}k_r \to \mathcal{E}$ by letting $f(n) = r^n$ on objects and letting $f(\tau)$ be τ regarded as a permutation of the set k_r^n of r^n letters on morphisms $\tau \in GL(n, k_r)$. Of course, f depends on the chosen isomorphism of sets from k_r^n to $1, 2, \ldots, r^n$. With the obvious lexicographic choice, f gives an exponential morphism of permutative categories

$$f: (\mathcal{GL}k_r, \oplus, 0, c) \to (\mathcal{E}, \otimes, 1, \tilde{c}).$$

Moreover, the composite morphism of $\mathcal{Q}$-spaces

$$(DS^0, \mu) = (B\mathcal{E}, \theta) \xrightarrow{Be} (B\mathcal{GL}k_r, \theta) \xrightarrow{Bf} (B\mathcal{E}, \xi) = (DS^0, \xi)$$

coincides with the exponential unit e_r defined in Definition 2.2 since $B(fe)$ sends 0 to 1 and 1 to a point in the component $\mathcal{Q}(r)/\Sigma_r$ of DS^0. This works equally well with $\mathcal{GL}k_r$ replaced by $\mathcal{O}k_r$.

Example 5.7. Let k be a field of characteristic $\neq 2$. $O(n, k)$ consists of the isometries with respect to the bilinear form B associated to the standard quadratic map $Q: k^n \to k$, $Q(x_1, \ldots, x_n) = \Sigma x_i^2$ [51, p. 84]. Recall from [51, p. 137] that the spinor norm $\nu: O(n, k) \to \dot{k}/(\dot{k})^2$ is defined by

$$\nu(\tau) = Q(z_1) \ldots Q(z_r) \text{ if } \tau = \tau_{z_1} \ldots \tau_{z_r} \in O(n, k).$$

Here $\tau_y(x) = x - [2B(x, y)/Q(y)]y$ for $x, y \in k^n$ with $y \neq 0$. Every τ is a product of such symmetries [51, p. 102] and, modulo squares, $\nu(\tau)$ is independent of the choice of factorization. If $y = e_i - e_j$, then τ_y permutes e_i and e_j and $\nu(\tau_y) = 2$. Now specialize to $k = k_3$. Then ν

takes values in Z_2 and $\nu(\sigma)\det(\sigma) = 1$ for $\sigma \in \Sigma_n \subset O(n, k_3)$. The subcategory $\mathcal{N}k_3$ of $\mathcal{O}k_3$ whose morphisms $n \to n$ are those $\tau \in O(n, k_3)$ such that $\nu(\tau)\det(\tau) = 1$ is a sub bipermutative category since ν and det are given by formulas of the same form on direct sums and tensor products. Again, Example 5.6 works equally well with $\mathcal{GL}k_3$ replaced by $\mathcal{N}k_3$.

We have only listed examples to which we shall refer in chapter VIII. As pointed out by Swan [unpublished], all of our examples, and many others, can be subsumed within a general framework of systems of groups $G(n)$ for $n \geq 0$ together with homomorphisms $\Sigma_n \to G(n)$, $G(m) \times G(n) \to G(m+n)$, and, for the bipermutative case, $G(m) \times G(n) \to G(mn)$ subject to the appropriate axioms. The following remarks, which apply to any such example, describe the action maps

$$\theta_p : \mathcal{Q}(p) \times_{\Sigma_p} BG(n)^p \to BG(pn) \quad \text{and} \quad \xi_p : \mathcal{Q}(p) \times_{\Sigma_p} BG(n)^p \to BG(n^p)$$

solely in terms of homomorphisms of groups. When p is a prime, the induced maps of mod p homology determine operations on $H_*(\coprod_{n \geq 0} BG(n); Z_p)$ [26, I], and the computation of these operations is thus reduced to the homological analysis of appropriate representations.

Remarks 5.8. Let $(\mathcal{A}, \oplus, 0, c)$ be a permutative category with objects $\{n \mid n \geq 0\}$ and with morphisms from n to n forming a topological group $G(n)$ which contains Σ_n. Recall that the wreath product $\Sigma_p \int G(n)$ is the semi-direct product of Σ_p and $G(n)^p$ determined by the evident action of Σ_p on $G(n)^p$. If we regard $\Sigma_p \int G(n)$ and $G(n)$ as categories with a single object, then $\Sigma_p \int G(n)$ is the orbit category $\tilde{\Sigma}_p \times_{\Sigma_p} G(n)^p$ of $\tilde{\Sigma}_p \times G(n)^p$. The functor $c_p : \tilde{\Sigma}_p \times G(n)^p \to G(pn)$ factors through the homomorphism $\Sigma_p \int G(n) \to G(pn)$ specified by

$$(\sigma; g_1, \dots, g_p) \to \sigma(n, \dots, n)(g_1 \oplus \dots \oplus g_p) = (g_{\sigma^{-1}(1)} \oplus \dots \oplus g_{\sigma^{-1}(p)})\sigma(n, \dots, n).$$

Application of the classifying space functor B thus gives the commutative diagram

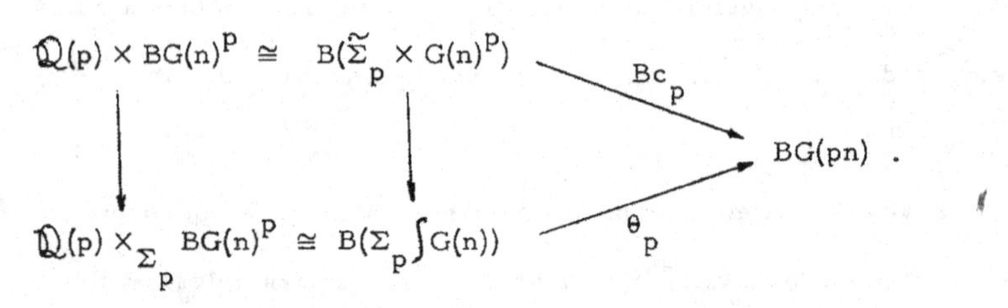

If, further, $(\mathcal{Q}, \oplus, 0, c, \otimes, 1, \tilde{c})$ is bipermutative, then the functor $\tilde{c}_p : \tilde{\Sigma}_p \times G(n)^p \to G(n)^p$ factors through the homomorphism $\Sigma_p \int G(n) \to G(n^p)$ specified by

$$(\sigma; g_1, \dots, g_p) \to \sigma\langle n, \dots, n\rangle(g_1 \otimes \dots \otimes g_p)$$

$$= (g_{\sigma^{-1}(1)} \otimes \dots \otimes g_{\sigma^{-1}(p)})\sigma\langle n, \dots, n\rangle$$

and application of B gives the commutative diagram

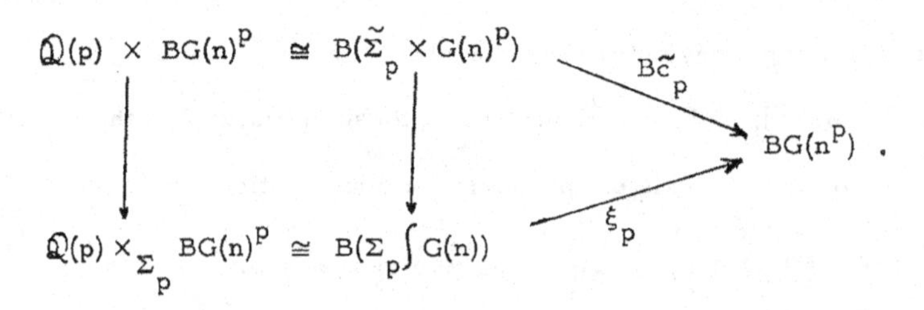

We use these remarks to determine $B\mathcal{O}Z$. The following observations are due to Z. Fiedorowicz.

Remarks 5.9. A moment's reflection will convince the reader that, for an integer valued matrix M, $MM^t = I$ if and only if each row and column of M has precisely one non-zero entry and that entry is ± 1. Indeed, the natural homomorphism $\Sigma_p \int O(1, Z) \to O(p, Z)$ is an isomorphism for all p. Abbreviate $\pi = O(1, Z)$ and regard $B\pi^+$, the union of $B\pi \simeq RP^\infty$ and a disjoint basepoint

0, to be the sub $\mathcal{Q}_0$-space $BO(0,Z) \amalg BO(1,Z)$ of $B\mathcal{O}Z$. $D(B\pi^+)$ is the free $(\mathcal{Q},\mathcal{Q})$-space generated by $B\pi^+$, and there results a map $D(B\pi^+) \rightarrow B\mathcal{O}Z$ of $(\mathcal{Q},\mathcal{Q})$-spaces. In fact, this map is just the identification (compare [45, 2.4] or [26, I. 5.7])

$$\amalg \theta_p : D(B\pi^+) \cong \amalg \mathcal{Q}(p) \times_{\Sigma_p} BO(1,Z)^p \longrightarrow \amalg BO(p,Z) = B\mathcal{O}Z.$$

The functor $e: \mathcal{E} \rightarrow \mathcal{O}Z$ gives rise under B to the injection $DS^0 \rightarrow D(B\pi^+)$ determined by the points 0 and 1 of $B\pi^+$.

VII. The recognition principle for E_∞ ring spaces

The zeroth space of a spectrum is an E_∞ space, and an E_∞ space determines a spectrum and therefore a cohomology theory. If ΓX denotes the zeroth space of the spectrum associated to an E_∞ space X, then there is a map $\iota : X \to \Gamma X$ which respects the E_∞ structure and is a group completion, in the sense that $\iota_*: H_*(X; k) \to H_*(\Gamma X; k)$ is a localization of the (Pontryagin) ring $H_*(X; k)$ at its submonoid $\pi_0 X$ for every commutative coefficient ring k. (See [46, §1] for a discussion of this definition; the letter Γ is chosen as a reminder of the group completion property and has nothing to do with the use of this letter in other theories of infinite loop spaces.) Here spectra are to be understood in the coordinate-free sense introduced in chapter II, and the results of [45 and 46] just summarized will be recast in terms of such spectra in section 3.

In chapter IV, Frank Quinn, Nigel Ray, and I introduced the notion of an E_∞ ring spectrum. In section 2, the zeroth space of such a spectrum will be shown to be an E_∞ ring space. The proof requires use of the little convex bodies operads $\mathcal{K}_V$ introduced in section 1; the essential feature of $\mathcal{K}_V$ is that the orthogonal group OV acts on it. In section 4, the spectrum determined by the additive E_∞ structure of an E_∞ ring space will be shown to be an E_∞ ring spectrum and it will be proven that, for an E_∞ ring space X, $\iota : X \to \Gamma X$ respects both E_∞ space structures. In effect, this means that the multiplicative E_∞ structure is preserved on passage from the additive E_∞ structure to its associated spectrum. As a special case of more general results, we shall see that if X is the disjoint union $\coprod K(\Sigma_j, 1)$, then ΓX is equivalent as an E_∞ ring space to QS^0. This result is a multiplicative elaboration of the Barratt-Quillen theorem [16;68;46 §3].

The component $\Gamma_1 X$ of the identity element of ΓX is a multiplicative E_∞ space. We prove in section 5 that, under mild hypotheses, the localization of $\Gamma_1 X$ at any submonoid M of the positive integers is equivalent as an infinite loop space to the component of the identity element of the zeroth space of the spectrum derived from the multiplicative E_∞ space structure on a certain subspace X_M of X. In other words, although $\Gamma_1 X$ is constructed by use of the additive E_∞ space structure on X, its localizations depend only on the multiplicative E_∞ space structure. In earlier approaches to multiplicative structures, only the localized infinite loop spaces were visible because there was no way to handle the additive and multiplicative structures in combination. In particular, the main result of Tornehave's paper [76], which describes localizations of SF in terms of the symmetric groups, will drop out as a special case by use of our version of the Barratt-Quillen theorem.

The essential results of this chapter were obtained in 1972 and presented in lectures during the winter of 1973. I mention this since at least one other author has since announced his intention of developing a similar theory.

§1. The little convex bodies operads

The little cubes operad $\mathcal{C}_n$ of [45, §4] played a canonical role in the passage from E_∞ spaces to spectra. Indeed, as explained in [45, p. 153-155], the geometry given by the action of any E_∞ operad on a space was automatically transformed into the little cubes geometry on the derived infinite loop space. We need a canonical E_∞ operad pair in order to obtain the analogous (but considerably more delicate) passage from E_∞ ring spaces to E_∞ ring spectra. From the definition of E_∞ ring spectra in chapter IV, it is clear that the linear isometries operad $\mathcal{L}$ of I.1.2 must be chosen as the canonical operad for the multiplicative structure. We require an operad $\mathcal{K}_\infty$ on which $\mathcal{L}$ acts and which can be used interchangeably with $\mathcal{C}_\infty$ in the additive, or one operad, theory.

Recall the definitions of $\mathcal{J}$ and $\mathcal{J}_*$ from I.1.1 and I.1.8. Let $\mathcal{J}^+$ and $\mathcal{J}_*^+$ denote their respective full subcategories of positive dimensional real inner product spaces. Ideally, we would like to construct a functor $\mathcal{K}$ from $\mathcal{J}^+$ to the category of operads such that application of $\mathcal{K}$ to R^n, $1 \leq n \leq \infty$, yields an operad $\mathcal{K}_n$ equivalent to $\mathcal{C}_n$. In fact, we shall have to settle for a good deal less. While $\mathcal{K}_\infty$ will be (weakly) equivalent to $\mathcal{C}_\infty$, the $\mathcal{K}_n$ for $n < \infty$ will not be (or at least will not be proven to be) equivalent to the $\mathcal{C}_n$. Moreover, in order to construct the functor $\mathcal{K}$ at all, we shall have to weaken the notion of operad and shall have to carefully examine the resulting geometric structures in order to make sure that the machinery of [45 and 46] still applies.

The difficulties can be explained quite simply. To carry out our original program, we would have to construct a space $\mathcal{E}V$ of embeddings $V \rightarrow V$ for each finite dimensional real inner product space V such that the following properties were satisfied:

(1) $fcf^{-1} \in \mathcal{E}W$ if $c \in \mathcal{E}V$ and $f \in \mathcal{J}(V, W)$, $\dim V = \dim W$.

(2) $c \times d \in \mathcal{E}(V \oplus W)$ if $c \in \mathcal{E}V$ and $d \in \mathcal{E}W$.

(3) $c \circ c' \in \mathcal{E}V$ if $c, c' \in \mathcal{E}V$.

(4) The space of j-tuples of elements of $\mathcal{E}V$ with pairwise disjoint images has the Σ_j-equivariant homotopy type of the configuration space $F(V, j)$ of j-tuples of distinct points of V.

I have been unable to construct such spaces $\mathcal{E}V$, and there is reason to believe that no such spaces of embeddings exist. We shall be forced to replace (4) by the following weaker condition

(4') $\mathcal{E}V$ is a contractible space.

Even then, the closure conditions (1), (2), and (3) appear to be incompatible, and we shall be forced to drop (3) altogether. However, we shall have to have spaces of composable sequences of embeddings which satisfy analogs of (1), (2), and (4'). With these considerations in mind, we proceed to our basic definitions.

<u>Definition 1.1.</u> Let V be a finite dimensional real inner product space. A little convex body in V is a topological embedding $c: V \to V$ such that the maps $c_t: V \to V$ specified by

$$c_t(x) = tx + (1-t)c(x) \quad \text{for } x \in V$$

are also embeddings for all $t \in I$. Since $c_{t,s} = c_{s+t-st}$, it follows that each c_t is again a little convex body. A sequence $(c_1, \ldots, c_q)$ of little convex bodies is said to be composable if $q = 1$ or, inductively, if $q > 1$ and

$$(c_{1,t}, \ldots, c_{i-1,t}, c_{i,t} \circ c_{i+1,t}, c_{i+2,t}, \ldots, c_{q,t}), \quad 1 \leq i < q,$$

is a composable sequence of little convex bodies for all $t \in I$. It follows that each $(c_{1,t}, \ldots, c_{q,t})$ is again composable and, by inductive use of the case $t = 0$, that all sequences obtained by composing some of the maps $c_{i,t}$ (in ordered blocks, with t fixed) are composable.

Although the definition evolved from convexity considerations, the term convex body is a misnomer: the image of a little convex body need not be convex.

Examples 1.2. (i) If $c: R \to R$ is an embedding which is an increasing function, then so is each c_t. Any sequence $(c_1, \ldots, c_r)$ of increasing embeddings $R \to R$ is composable.

(ii) If $c: V \to V$ is a little convex body and $f \in \mathcal{I}(V, W)$, $\dim V = \dim W$, then $(fcf^{-1})_t = fc_t f^{-1}$. If $(c_1, \ldots, c_q)$ is a composable sequence of little convex bodies in V, then $(fc_1 f^{-1}, \ldots, fc_q f^{-1})$ is a composable sequence of little convex bodies in W.

(iii) If $c: V \to V$ and $d: W \to W$ are little convex bodies, then $(c \times d)_t = c_t \times d_t$. If $(c_1, \ldots, c_q)$ and $(d_1, \ldots, d_q)$ are composable sequences of little convex bodies in V and in W, then $(c_1 \times d_1, \ldots, c_q \times d_q)$ is a composable sequence of little convex bodies in $V \oplus W$.

For rigor, we should at this point define the notion of a "partial operad". However, to avoid excess verbiage, we prefer to be informal. We agree to continue to use the term operad for structures specified as in Definition VI.1.2, but with the structural maps γ defined only on specified subspaces of the spaces $\mathcal{C}(k) \times \mathcal{C}(j_1) \times \ldots \times \mathcal{C}(j_k)$, with the two-fold iterates of the γ (as in VI.1.2(a)) defined only on specified subspaces of the spaces

$$\mathcal{C}(k) \times \mathcal{C}(j_1) \times \ldots \times \mathcal{C}(j_k) \times \mathcal{C}(i_1) \times \ldots \times \mathcal{C}(i_{j_1 + \ldots + j_k}),$$

and so forth. The only examples will be the little convex bodies operads (and their products with honest operads), where the γ will be obtained by composition and their domains will be specified by allowing only composable sequences (in the sense of Definition 1.1) to be composed.

Definition 1.3. Let V be a finite dimensional real inner product space. Define the little convex bodies operad $\mathcal{K}_V$ of V as follows. Let $\mathcal{K}_V(j)$ be the set of those j-tuples $<c_1, \ldots, c_j>$ of little convex bodies such that the images of the c_r are pairwise disjoint.

Let ${}^jV$ denote the disjoint union of j copies of V, regard $<c_1, \ldots, c_j>$ as a map ${}^jV \to V$, and topologize $\mathcal{K}_V(j)$ as a subspace of the space of all continuous functions ${}^jV \to V$. (Regard $\mathcal{K}_V(0) = < >$ as the unique "embedding" of the empty set in V.) The requisite data are specified by

(a) $\gamma(c; d_1, \ldots, d_k) = c \circ (d_1 + \ldots + d_k): {}^{j_1 + \ldots + j_k}V = {}^{j_1}V + \ldots + {}^{j_k}V \to V$ for those $c = <c_1, \ldots, c_k> \in \mathcal{K}_V(k)$ and $d_r = <d_{r,1}, \ldots, d_{r,j_r}> \in \mathcal{K}_V(j_r)$ such that each pair $(c_r, d_{r,s})$ is a composable sequence;

(b) $1 \in \mathcal{K}_V(1)$ is the identity function; and

(c) $<c_1, \ldots, c_j> \sigma = <c_{\sigma(1)}, \ldots, c_{\sigma(j)}>$ for $\sigma \in \Sigma_j$.

Clearly the action of Σ_j on $\mathcal{K}_V(j)$ is free. For $f \in \mathcal{J}(V, W)$, $\dim V = \dim W$, define a morphism of operads $\mathcal{K}_f: \mathcal{K}_V \to \mathcal{K}_W$ by $c \to fcf^{-1}$ on little convex bodies. Then $\mathcal{K}$ is a functor from $\mathcal{J}_*^+$ to the category of (partial) operads. For finite dimensional inner product spaces V and W, define a morphism of operads $\sigma: \mathcal{K}_V \to \mathcal{K}_{V \oplus W}$ by $c \to c \times 1$ on little convex bodies. By passage to limits, precisely as in the proof of I. 6. 9, $\mathcal{K}$ extends to a functor from $\mathcal{J}^+$ to the category of operads. For $1 \leq n \leq \infty$, define $\mathcal{K}_n = \mathcal{K}_V$ where $V = R^n$.

Recall from [45, §4] that a little n-cube is a map $c: J^n \to J^n$, where $J = (0, 1)$, which is a linear embedding with parallel axes (i. e., a product of n increasing linear embeddings $J \to J$). The little n-cubes operad $\mathcal{C}_n$ is defined in precisely the same way that the $\mathcal{K}_V$ were, but here the γ are everywhere defined. Examples 1. 2 imply the following pair of results.

Lemma 1.4. Let $\alpha: J \to R$ be an increasing homeomorphism and let $f \in \mathcal{J}(R^n, V)$, $\dim V = n$. Then the maps $\mathcal{C}_n(j) \to \mathcal{K}_V(j)$ specified by sending a little n-cube c to the little convex body $f\alpha^n c(\alpha^{-1})^n f^{-1}$ define a morphism of operads $\mathcal{C}_n \to \mathcal{K}_V$. Thus $\mathcal{K}_V$ contains a copy of $\mathcal{C}_n$ for each such pair (α, f).

Lemma 1.5. Fix an increasing homeomorphism $\alpha: J \to R$ and let $i_n: \mathcal{C}_n \to \mathcal{K}_n$ be the morphism of operads specified by $c \to \alpha^n c(\alpha^{-1})^n$ on little n-cubes. Then $i_{n+1} \circ \sigma = \sigma \circ i_n : \mathcal{C}_n \to \mathcal{K}_{n+1}$ and the i_n induce a morphism of operads $i_\infty : \mathcal{C}_\infty \to \mathcal{K}_\infty$ by passage to limits.

Assume that $\alpha(1/2) = 0$ and let $g: \mathcal{C}_n(j) \to F(J^n, j)$ and $f: \mathcal{K}_V(j) \to F(R^n, j)$ be the maps specified by sending a little cube c to its center point $c(1/2, \ldots, 1/2)$ and a little convex body c to its center point $c(0)$. Define a homeomorphism $j_n: F(J^n, j) \to F(R^n, j)$ of configuration spaces by $x \to \alpha^n(x)$ on points $x \in J^n$. Then the following Σ_j-equivariant diagram is commutative:

$$\begin{array}{ccc} \mathcal{C}_n(j) & \xrightarrow{i_n} & \mathcal{K}_n(j) \\ \downarrow g & & \downarrow f \\ F(J^n, j) & \xrightarrow{j_n} & F(R^n, j) \end{array}$$

By [45,4.8], g is a Σ_j-equivariant homotopy equivalence. Thus, up to homotopy, $\mathcal{C}_n(j)$ is a Σ_j-equivariant retract of $\mathcal{K}_n(j)$. I have not been able to prove that i_n is actually a Σ_j-equivariant homotopy equivalence (although this could perhaps be arranged at the price of a more complicated notion of little convex body).

By the very definition of composable sequences of little convex bodies in V, the deformation specified by $h_t(c) = c_t$ contracts the space of such sequences of length q to the identity sequence $(1, \ldots, 1)$. In particular, it follows that $\sigma: \mathcal{K}_V(j) \to \mathcal{K}_{V\oplus W}(j)$ is null homotopic for all V and W. Indeed, the requisite deformation k is specified by choosing any point $<d_1, \ldots, d_j> \in \mathcal{K}_W(j)$ and defining

$$k_t(c_1 \times 1, \ldots, c_j \times 1) = \begin{cases} <c_1 \times d_{1,1-2t}, \ldots, c_j \times d_{j,1-2t}> , & 0 \leq t \leq 1/2 \\ <c_{1,2t} \times d_1, \ldots, c_{j,2t} \times d_j> , & 1/2 \leq t \leq 1 . \end{cases}$$

The point is that the disjoint image requirement is satisfied on the first half of the deformation because $<c_1, \ldots, c_j> \in \mathcal{K}_V(j)$ and on the second half because $<d_1, \ldots, d_j> \in \mathcal{K}_W(j)$. Therefore $\pi_* \mathcal{K}_\infty(j) = 0$. (In the case of little cubes, this argument is due to Boardman and Vogt [20, p. 65].) By the same argument, the product of $\sigma: \mathcal{K}_V(k) \to \mathcal{K}_{V\oplus W}(k)$ and the $\sigma: \mathcal{K}_V(j_r) \to \mathcal{K}_{V\oplus W}(j_r)$ for $1 \leq r \leq k$ restricts to a null homotopic map from the domain of γ (for $\mathcal{K}_V$) to the domain of γ (for $\mathcal{K}_{V\oplus W}$), hence the domain of γ for $\mathcal{K}_\infty$ has trivial homotopy groups, and similarly for the domains of all iterates of the maps γ. It would follow that $\mathcal{K}_\infty(j)$ was contractible if we knew either that $\mathcal{K}_\infty(j)$ had the homotopy type of a CW-complex or that each $\sigma: \mathcal{K}_n(j) \to \mathcal{K}_{n+1}(j)$ was a cofibration (and similarly for the domains of the maps γ and their iterates). I have not verified either assertion. However, we need not let this difficulty detain us since the conclusions we wish to derive from E_∞ space structures can be phrased homologically, so that it is harmless to require of a partial E_∞ operad $\mathcal{C}$ only that the $\mathcal{C}(j)$ and the domains of the γ and their iterates have trivial homology groups. Thus $i_\infty: \mathcal{C}_\infty \to \mathcal{K}_\infty$ may be regarded as a weak equivalence of (partial) E_∞ operads.

§2. The canonical E_∞ operad pair

We first show that just as $\mathcal{C}_n$ acts naturally on n-fold loop spaces [45, 5.1] so $\mathcal{K}_V$ acts naturally on "V-fold" loop spaces. We then discuss the "partial monads" K_V to which the partial operads $\mathcal{K}_V$ give rise. Finally, we show that $\mathcal{L}$ acts on $\mathcal{K}_\infty$, so that K_∞ restricts to a (partial) monad in $\mathcal{L}[\mathcal{T}_e]$ (as in VI.1.9), and prove that the zeroth space of any E_∞ ring spectrum is an E_∞ ring space.

We should begin by defining the notion of an action by a (partial) operad on a space, but shall instead leave the requisite modification of Definition VI.1.3 to the reader. We continue to use the term E_∞ space for a space with an action by a (partial) E_∞ operad. With the appropriate definition, the following result is trivial to verify. We agree to fix tV as the one-point compactification of V, so that

$$\Omega^V X = F(tV, X) \quad\text{and}\quad \Sigma^V X = X \wedge tV.$$

We agree to identify tR^n with $tJ^n = I^n/\partial I^n$ via the homeomorphism $t(\alpha^{-1})^n$, where $\alpha: J \to R$ is as in Lemma 1.5.

Theorem 2.1. For $X \in \mathcal{T}$ and $V \in \mathcal{J}_*^+$, let $\theta_{V,j}: \mathcal{K}_V(j) \times (\Omega^V X)^j \to \Omega^V X$ be the map specified by

$$\theta_{V,j}(\langle c_1, \ldots, c_j\rangle, y_1, \ldots, y_j)(b) = \begin{cases} y_r(a) & \text{if } c_r(a) = b \in V \\ * & \text{if } b \notin \bigcup_{r=1}^{j} \operatorname{Im} c_r \end{cases}$$

Then the $\theta_{V,j}$ define an action θ_V of $\mathcal{K}_V$ on $\Omega^V X$. Under the natural identification $\Omega^V \Omega^W X = \Omega^{V\oplus W} X$, the action θ_V coincides with the composite of $\theta_{V\oplus W}$ and $\sigma: \mathcal{K}_V \to \mathcal{K}_{V\oplus W}$. For $E \in \mathcal{S}$, the actions θ_V of $\mathcal{K}_V$ on $E_0 \cong \Omega^V EV$ for $V \subset R^\infty$ define an action θ_∞ of $\mathcal{K}_\infty$ on E_0. Under the identification $\Omega^n X = \Omega^V X$, $V = R^n$, the action of $\mathcal{C}_n$ on $\Omega^n X$ coincides with the composite of $i_n: \mathcal{C}_n \to \mathcal{K}_n$ and the action θ_n of $\mathcal{K}_n$

on $\Omega^n X$, and similarly for the action of $\mathcal{C}_\infty$ on E_0.

Define a space $K_V X$ for each space X by letting

$K_V X = \coprod \mathcal{K}_V(j) \times_{\Sigma_j} X^j/(\approx)$, where

$(<c_1,\ldots,c_j>, x_1,\ldots,x_j) \approx (<c_1,\ldots,c_{i-1},c_{i+1},\ldots,c_j>, x_1,\ldots,x_{i-1},x_{i+1},\ldots,x_j)$

whenever $x_i = *$ (compare [45, 2.4]). Define $\eta: X \to K_V X$ by $\eta(x) = [<1>, x]$. We would like to define $\mu: K_V K_V X \to K_V X$ by

$$\mu[c; [d_1, y_1], \ldots, [d_k, y_k]] = [\gamma(c; d_1, \ldots, d_k), y_1, \ldots, y_k]$$

for $c \in \mathcal{K}_V(k)$, $d_r \in \mathcal{K}_V(j_r)$, and $y_r \in X^{j_r}$. We cannot do so since γ is not defined on all of $\mathcal{K}_V(k) \times \mathcal{K}_V(j_1) \times \ldots \times \mathcal{K}_V(j_r)$. We should therefore define the notion of a partial monad C, with structural maps $\mu: C^{(2)}X \to CX$ for a suitable subspace $C^{(2)}X$ of CCX. The various (n-1)-fold iterates of μ would have to be defined (and equal) on specified subspaces $C^{(n)}X$ of $C^n X$. (More precisely, $C^{(n)}$ would be required to be a subfunctor of C^n.) With the proper formal definitions, a partial operad $\mathcal{C}$ gives rise to a partial monad C in such a way that the notions of an action by $\mathcal{C}$ and of an action by C are equivalent. Here again, we leave the pedantic details to the reader and continue to use the terms monad and C-space for the more general concepts. The only examples will be the monads C derived from operads of the form $\mathcal{C}' \times \mathcal{K}_V$ (where $\mathcal{C}'$ is an honest operad). Here the subspaces $C^{(n)}X$ of $C^n X$ are obvious: all n-fold sequences of little convex bodies which would be composed under the iterates of μ are required to be composable in the sense of Definition 1.1.

We have an action $\theta_V: K_V \Omega^V X \to \Omega^V X$ for any space X.

<u>Theorem 2.2.</u> For $X \in \mathcal{T}$ and $V \in \mathcal{J}_*^+$, define $\alpha_V: K_V X \to \Omega^V \Sigma^V X$ to be the composite

$$K_V X \xrightarrow{K_V \eta} K_V \Omega^V \Sigma^V X \xrightarrow{\theta_V} \Omega^V \Sigma^V X,$$

where $\eta: X \to \Omega^V\Sigma^V X$ is the natural inclusion, and define $\alpha_\infty: K_\infty X \to QX$ by passage to limits over $V \subset R^\infty$. Then $\alpha_V: K_V \to \Omega^V\Sigma^V$ and $\alpha_\infty: K_\infty \to Q$ are morphisms of monads in $\mathcal{J}$ and the actions of K_V on $\Omega^V X$ and of K_∞ on E_0 for $E \in \mathcal{S}$ are induced by pullback along α_V and α_∞ from the actions of $\Omega^V\Sigma^V$ and of Q. Moreover, α_∞ is a weak homotopy equivalence if X is connected and is a group completion in general.

Proof. The monads $\Omega^V\Sigma^V$ and Q are defined as in [45, p. 17 and 46], and the first statement holds by slight elaborations of the purely formal diagram chases in the proof of [45, 5.2]; moreover, with $\alpha_n = \alpha_V$ for $V = R^n$, $\alpha_n i_n: C_n \to \Omega^n\Sigma^n$ coincides with the morphism of monads derived in the cited result. By [46, A. 2(i)], $i_\infty: C_\infty X \to K_\infty X$ is a homology isomorphism for any space X. Now the last statement for X connected is given by [45, 6.1], while the general case is proven by explicit homological calculation in [26, I§ 5].

We next exploit the fact that $\mathcal{K}_\infty$ is obtained by use of the functor $\mathcal{K}$ (from $\mathcal{I}^+$ to operads) to specify an action of $\mathcal{L}$ on $\mathcal{K}_\infty$. Yet again, we can and must first generalize all of the definitions and results VI. 1.6-1.11 so as to allow the additive operad of an operad pair to be a partial operad. The multiplicative operad will be required to be honest. We omit the details, and we continue to use the term E_∞ ring space for a space with an action by a (partial) E_∞ operad pair.

Lemma 2.3. $(\mathcal{K}_\infty, \mathcal{L})$ is an E_∞ operad pair with respect to the action maps λ specified on $g \in \mathcal{L}(k)$ and $c_r \in \mathcal{K}_\infty(j_r)$ by

$$\lambda(g; c_1, \ldots, c_k) = \langle \underset{I \in S(j_1, \ldots, j_k)}{\times} \mathcal{K}_g(c_I) \rangle \in \mathcal{K}_\infty(j_1 \cdots j_k)$$

where, if $c_r = \langle c_{r1}, \ldots, c_{rj_r} \rangle$ and $I = \{i_1, \ldots, i_k\}$, then c_I is the little convex body $c_{1i_1} \times \ldots \times c_{ki_k}$ in $(R^\infty)^k$.

Here $\mathcal{K}_g(c_I)$ is $g c_I g^{-1}$ on finite dimensional subspaces gW of R^∞ and is the identity on the orthogonal complement of gW whenever c_I is the identity on the orthogonal complement of $W \subset (R^\infty)^k$. The verifications of the identities specified in VI.1.6 are tedious, but elementary.

If $\rho: \mathcal{G} \to \mathcal{L}$ is a morphism of E_∞ operads, then, by IV.1.9, Q is a monad in $\mathcal{G}[\mathcal{T}_e]$. As explained in VI §1, if $\mathcal{G}$ acts on $\mathcal{C}$, then C is a (partial) monad in $\mathcal{G}[\mathcal{T}_e]$. We have the following consistency statement, which implies that the zeroth space of any E_∞ ring spectrum is an E_∞ ring space.

Theorem 2.4. Let $(\pi, \rho): (\mathcal{C}, \mathcal{G}) \to (\mathcal{K}_\infty, \mathcal{L})$ be a morphism of E_∞ operad pairs. Then the morphisms $\pi: C \to K_\infty$ and $\alpha_\infty: K_\infty \to Q$ of monads in $\mathcal{T}$ restrict to morphisms of monads in $\mathcal{G}[\mathcal{T}_e]$. If E is a $\mathcal{G}$-spectrum, then its zeroth space E_0 is a $(\mathcal{C}, \mathcal{G})$-space by pullback of its Q-action $QE_0 \to E_0$ along $\alpha_\infty \pi$.

Proof. By IV.1.9, the second statement will follow from the first. Clearly $\mathcal{G}$ acts on $\mathcal{K}_\infty$ via

$$\lambda(g; c_1, \ldots, c_k) = \lambda(\rho g; c_1, \ldots, c_k)$$

and $(\pi, 1): (\mathcal{C}, \mathcal{G}) \to (\mathcal{K}_\infty, \mathcal{G})$ is a morphism of operad pairs. Let $X \in \mathcal{G}[\mathcal{T}_e]$. $\pi: CX \to K_\infty X$ is a morphism of $\mathcal{G}_0$-spaces in view of VI.1.8. Since $K_\infty \eta: K_\infty X \to K_\infty QX$ is a morphism of $\mathcal{G}_0$-spaces, because η is so, $\alpha_\infty: K_\infty X \to QX$ will be a morphism of $\mathcal{G}_0$-spaces provided that $\theta_\infty: K_\infty QX \to QX$ is so. Thus we must verify that the following diagrams commute:

$$\begin{array}{ccc}
\mathcal{G}(k) \times \mathcal{K}_\infty(j_1) \times (QX)^{j_1} \times \ldots \times \mathcal{K}_\infty(j_k) \times (QX)^{j_k} & \xrightarrow{1 \times \theta_{\infty, j_1} \times \ldots \times \theta_{\infty, j_k}} & \mathcal{G}(k) \times (QX)^k \\
\downarrow \xi_k & & \downarrow \xi_k \\
\mathcal{K}_\infty(j_1 \cdots j_k) \times (QX)^{j_1 \cdots j_k} & \xrightarrow{\theta_{\infty, j_1 \cdots j_k}} & QX
\end{array}$$

ξ_k on the left is specified in VI.1.8. Think of points of QX as maps $S^\infty \to X \wedge S^\infty$, where $S^\infty = tR^\infty$. For $c = \langle c_1, \ldots, c_j \rangle \in \mathcal{K}_\infty(j)$ and $y = (y_1, \ldots, y_j) \in (QX)^j$, $\theta_{\infty,j}(c, y)$ is then the composite

$$S^\infty \xrightarrow{\bar{c}} {}^jS^\infty \xrightarrow{y_1 \vee \ldots \vee y_j} {}^j(X \wedge S^\infty) \xrightarrow{p} X \wedge S^\infty ,$$

where ${}^jY$ denotes the wedge of j copies of Y, $\bar{c}$ is the pinch map specified by $\bar{c}(b) = *$ unless $b = c_r(a)$ for some r and a, when $\bar{c}(b) = a$ in the rth copy of S^∞, and p is the evident folding map. For $g \in \mathcal{G}(k)$ and $z = (z_1, \ldots, z_k) \in (QX)^k$, $\xi_k(g, z)$ is that map which makes the following diagram commute

$$\begin{array}{ccccc} S^\infty & \xleftarrow{\rho g} & S^\infty \wedge \ldots \wedge S^\infty & \xrightarrow{z_1 \wedge \ldots \wedge z_k} & X \wedge S^\infty \wedge \ldots \wedge X \wedge S^\infty \\ \downarrow{\scriptstyle \xi_k(g,z)} & & & & \wr\Vert \\ X \wedge S^\infty & \xleftarrow{1 \wedge \rho g} & X \wedge S^\infty \wedge \ldots \wedge S^\infty & \xleftarrow{\xi_k(g) \wedge 1} & X \wedge \ldots \wedge X \wedge S^\infty \wedge \ldots \wedge S^\infty \end{array}$$

and which, on the orthogonal complement of $(\rho g)t(V_1 \oplus \ldots \oplus V_k)$ where $V_r \subset R^\infty$ is a finite-dimensional subspace such that z_r is in the image of $\Omega^{V_r}\Sigma^{V_r}X$, has constant coordinate in X and the identity map as coordinate in S^∞. Now the desired commutativity is easily verified by direct computation, the point being that the smash products used in the definition of ξ_k distribute over the wedge sums used in the definition of the $\theta_{\infty j}$.

§3. The one operad recognition principle.

We translate the one operad, additive, recognition principle of [45, §14 and 46, §2] into the language of coordinate-free spectra. It simplifies slightly in the process since the construction of a prespectrum and the passage from a prespectrum to a spectrum were awkwardly combined in the earlier versions. It also complexifies slightly since we must take account of the distinction between partial and honest operads and their actions. However, the basic constructions and the bulk of the proofs remain unchanged and will not be repeated here.

Let $\mathcal{C}'$ be a locally contractible (honest) operad, for example $\mathcal{N}$ or an E_∞ operad. Define $\mathcal{C} = \mathcal{C}' \times \mathcal{K}_\infty$ [45, 3.8] and observe that $\mathcal{C}$ is a (partial) E_∞ operad (that is, Σ_j acts freely on $\mathcal{C}(j)$ and $\mathcal{C}(j)$ and the domains of the structural maps γ and their iterates have trivial homology groups). Let $\pi: \mathcal{C} \to \mathcal{K}_\infty$ and $\psi: \mathcal{C} \to \mathcal{C}'$ be the projections.

Let (X, θ) be a $\mathcal{C}$-space. For technical reasons, we assume once and for all that X is of one of the following three types (which certainly include all examples of any interest).

(1) The action of $\mathcal{C}$ on X is obtained by pullback along $\psi: \mathcal{C} \to \mathcal{C}'$ from an action of $\mathcal{C}'$ on X.

(2) X is E_0 regarded as a C-space by pullback of its Q-space structure along $\alpha_\infty \pi$, where $E \in \mathcal{S}$.

(3) X is CY regarded as a C-space via the structural map γ of the (partial) monad C, where $Y \in \mathcal{J}$.

In (1) and (2), the domain of θ is CX; in (3), the domain of θ is the domain $C^{(2)}Y$ of μ.

For each finite dimensional sub inner product space V of R^∞, define $\mathcal{C}_V = \mathcal{C}' \times \mathcal{K}_V$ and let C_V be the monad in $\mathcal{J}$ associated to $\mathcal{C}_V$. Recall, and generalize to the context of partial monads, the notion of a (right) action of a monad on a functor [45, 9.4]. By [45, 9.5], the adjoint of $\alpha_V: K_V \to \Omega^V \Sigma^V$ gives an action of K^V on the functor Σ^V. By pullback along $\pi: C_V \to K_V$, we obtain an action β_V of C_V on Σ^V.

The basic geometric construction of [45] is the two-sided bar construction [45, 9.6 and 11.1]

$$B(F,C,X) = |B_*(F,C,X)|,$$

where C is a monad, X is a C-space, and F is a C-functor. Here $|\ |$ denotes the geometric realization functor from simplicial spaces to spaces, and the space $B_q(F,C,X)$ of q-simplices is FC^qX where C^q denotes the q-fold iterate of C. This construction generalizes readily to the context of partial monads and their actions. In practice, due to X being of one of the three types specified above and to the definition of C and the C_V in terms of $\mathcal{C}'$, $\mathcal{K}_\infty$, and the $\mathcal{K}_V$, there will always be obvious subspaces $B_q(F,C,X)$ of FC^qX in sight so that the appropriate faces and degeneracies are defined. Indeed, this will simply amount to the requirement that precisely the composable sequences of little convex bodies (in the sense of Definition 1.1) are allowed to be composed.

We may thus define a space $(TX)(V)$ by

$$(4) \qquad (TX)(V) = B(\Sigma^V, C_V, X) \equiv |B_*(\Sigma^V, C_V, X)| .$$

By convention, when $V = \{0\}$, Σ^V, Ω^V, K_V, and C_V are all the identity functor on $\mathcal{J}$ and the α_V and β_V are identity maps. Thus the zeroth space $T_0X = (TX)\{0\}$ is just X.

For an orthogonal pair of finite dimensional subspaces V and W of R^∞, the morphism of operads $\mathcal{K}_V \to \mathcal{K}_{V+W}$ induces a morphism of operads $\mathcal{C}_V \to \mathcal{C}_{V+W}$. With the first equality given by [45, 9.7 and 12.1], we therefore obtain an inclusion

$$(5) \qquad \sigma: \Sigma^W B(\Sigma^V, C_V, X) = B(\Sigma^{V+W}, C_V, X) \to B(\Sigma^{V+W}, C_{V+W}, X) .$$

We would like to say that (TX, σ) gives a prespectrum, as defined in II.1.1. For this, T must be appropriately defined on isometries

$f: V \to V'$ for subspaces V and V' of R^∞ of the same finite dimension. It is easily verified by separate arguments in the cases (1), (2), and (3) that there are maps $\xi(f): X \to X$ such that if $\Sigma^f: \Sigma^V \to \Sigma^{V'}$ is defined in the obvious way and if $\mathcal{C}_f = 1 \times \mathcal{K}_f: \mathcal{C}_V \to \mathcal{C}_{V'}$, then maps $(TX)(f)$ as required can be specified by

(6) $\quad (TX)(f) = B(\Sigma^f, C_f, \xi(f)): B(\Sigma^V, C_V, X) \to B(\Sigma^{V'}, C_{V'}, X)$.

We omit the details since the requisite maps $\xi(f)$ will appear most naturally in the two operad theory (as use of the letter ξ would suggest) and since, as explained in II. 1. 10, the $(TX)(f)$ in any case play no essential role.

Thus (TX, σ) is a prespectrum. Consider the spectrum $\Omega^\infty TX$ and the natural map $\iota: TX \to \Omega^\infty TX$ of prespectra given by II. 1. 4 and II. 1. 5. Let ΓX, or $\Gamma(X, \theta)$ when necessary for clarity, denote the zeroth space of $\Omega^\infty TX$. The crux of the recognition principle is the analysis of the zeroth map $\iota: X \to \Gamma X$.

<u>Theorem 3.1</u>. Consider the following diagram:

$$\begin{array}{ccc} B(C, C, X) & \xrightarrow{B(\alpha_\infty \pi, 1, 1)} & B(Q, C, X) \\ \varepsilon(\theta) \downarrow \uparrow \tau(\eta) & & \downarrow \gamma^\infty \\ X & \xrightarrow{\iota} & \Gamma X \end{array}$$

(i) $\quad \varepsilon(\theta)$ is a strong deformation retraction with right inverse $\tau(\eta)$;

(ii) $\quad B(\alpha_\infty \pi, 1, 1)$ is a group completion and is therefore a weak homotopy equivalence if X is grouplike (i.e., if $\pi_0 X$ is a group);

(iii) $\quad \gamma^\infty$ is a weak homotopy equivalence;

(iv) $\quad \iota = \gamma^\infty \circ B(\alpha_\infty \pi, 1, 1) \circ \tau(\eta)$, hence ι is a group completion.

Moreover, $\varepsilon(\theta)$, $B(\alpha_\infty \pi, 1, 1)$, and γ^∞ (but not $\tau(\eta)$ and ι) are maps of $\mathcal{C}$-spaces.

Proof. Formal results from [45] apply equally well in the context of partial monads as in that of monads, and [45, 9.2, 9.8 and 11.10] imply (i). Results in [45] which apply to general simplicial spaces also apply equally well here, and [45, 12.3 and 14.4 (iii)] imply (iii). The first part of (iv) is a trivial calculation (compare [45, 14.4 (iv)]), and the last statement is proven by a slight elaboration, necessitated by our partial structures, of the proofs of [45, 12.2 and 12.4]. It remains to prove (ii). Here we shall have to use the infinite little cubes operad $\mathcal{C}_\infty$, and we rewrite $C = C' \times K_\infty$ and let $C' \times C_\infty$ be the monad associated to $\mathcal{C}' \times \mathcal{C}_\infty$. By use of $i_\infty : \mathcal{C}_\infty \to \mathcal{K}_\infty$, we obtain the commutative diagram

$$\begin{array}{ccccc} & & B(C' \times C_\infty, C' \times C_\infty, X) & \xrightarrow{B(\alpha_\infty i_\infty \pi, 1, 1)} & B(Q, C' \times C_\infty, X), \\ & \overset{\varepsilon(\theta)}{\swarrow} \simeq & \Big\downarrow B(1 \times i_\infty, 1 \times i_\infty, 1) & & \Big\downarrow B(1, 1 \times i_\infty, 1) \\ X & & & & \\ & \underset{\varepsilon(\theta)}{\nwarrow} \simeq & B(C' \times K_\infty, C' \times K_\infty, X) & \xrightarrow{B(\alpha_\infty \pi, 1, 1)} & B(Q, C' \times K_\infty, X) \end{array}$$

where X is regarded as a $C' \times C_\infty$-space by pullback along $1 \times i_\infty$. By the left triangle, $B(1 \times i_\infty, 1 \times i_\infty, 1)$ is a homotopy equivalence. By [46, 2.3], but with VI. 2.7 (iv) substituted for [46, 2.1] in the proof given there, the top arrow $B(\alpha_\infty i_\infty \pi, 1, 1)$ is a group completion. It therefore suffices to prove that $B(1, 1 \times i_\infty, 1)$ induces an isomorphism on homology. By [46, A.4], it suffices to show that each

$$B_q(1, 1 \times i_\infty, 1) : B_q(Q, C' \times C_\infty, X) \to B_q(Q, C' \times K_\infty, X)$$

induces an isomorphism on homology. $B_q(Q, C' \times C_\infty, X)$ and $B_q(Q, C' \times K_\infty, X)$ are obtained by application of the functor Q to $(C' \times C_\infty)^q(X)$ and to $(C' \times K_\infty)^{[q]}(X)$, where the latter is the appropriate domain space, namely $(C' \times K_\infty)^{(q)}(X)$ in cases (1) and (2) and

$(C' \times K_\infty)^{(q+1)}(Y)$ in case (3). By [26,I§4], it suffices to show that

$$(C' \times C_\infty)^q(X) \to (C' \times K_\infty)^{[q]}(X)$$

induces an isomorphism on homology. Typical points of both sides have coordinates in various of the spaces of the relevant operads and in X or Y. We may filter by the number of coordinates in X or Y. The successive quotients may be thought of as generalized equivariant half-smash products [45, 2.5], and the map induced on such spaces by any morphism of (partial) E_∞ operads is a homology isomorphism. Indeed, the shuffle map shows that the homology of such spaces depends only on the chains of X or Y and the chains of the operad coordinate. The latter chains are acyclic and free over the appropriate configuration of symmetric groups, hence standard techniques of homological algebra apply to yield the conclusion.

The following pair of results show that $\Omega^\infty TX$ gives the "right" spectrum in cases (2) and (3).

Proposition 3.2. Let $E \in \mathcal{S}$. Then the maps

$$\varepsilon(\phi_V): (TE_0)(V) = B(\Sigma^V, C_V, E_0) \to EV,$$

where $\phi_V: \Sigma^V E_0 \cong \Sigma^V \Omega^V EV \to EV$ is the evaluation map, define a natural map $\omega: TE_0 \to \nu E$ of prespectra. The unique map $\tilde{\omega}: \Omega^\infty TE_0 \to E$ of spectra such that $(\nu\tilde{\omega})\iota = \omega$ induces an isomorphism on π_i for all $i \geq 0$.

Proof. $\varepsilon(\phi_V)$ is defined in [45, 9.2 and p. 126], and the first statement is an easy verification from (4) through (6) and the definition, II. 1.1, of prespectra. Since $\omega: E_0 = (TE_0)\{0\} \to E_0$ is the identity map and E_0 is grouplike, (iv) of the theorem implies that $\tilde{\omega}: \Gamma E_0 \to E_0$ is a weak homotopy equivalence. The second statement follows.

The proposition implies that ω becomes an isomorphism in the stable homotopy category $H\mathcal{S}$ if E is connective. In other words, E (and thus the cohomology theory it determines) can be recovered from the underlying $\mathcal{C}$-space E_0.

Proposition 3.3. For $Y \in \mathcal{J}$, the composite map of spectra

$$\Omega^{\infty}TCY \xrightarrow{\Omega^{\infty}T(\alpha_{\infty}\pi)} \Omega^{\infty}TQY \xrightarrow{\tilde{\omega}} Q_{\infty}Y = \Omega^{\infty}\Sigma^{\infty}Y$$

is a strong deformation retraction.

Proof. Recall that $(\Sigma^{\infty}Y)(V) = \Sigma^{V}Y$. The maps

$$\varepsilon(\beta_V) : B(\Sigma^V, C_V, C_VY) \rightarrow \Sigma^V Y$$

are strong deformation retractions by [45, 9.9 and 11.10]. With the evident maps $\sigma = B(1, \sigma, \sigma)$ and $B(\Sigma^f, C_f, C_f)$ as in (5) and (6), the $B(\Sigma^V, C_V, C_VY)$ are the spaces of a prespectrum $T'CY$, and the maps $\varepsilon(\beta_V)$ define a strong deformation retraction of prespectra. There is an obvious inclusion of $T'CY$ in TCY and, since passage to spectra is a limit process over $V \subset R^{\infty}$, this inclusion becomes an isomorphism upon application of the functor Ω^{∞}. The resulting deformation retraction $\Omega^{\infty}TCY \rightarrow Q_{\infty}Y$ is the specified composite since β_V is the composite

$$\Sigma^V C_V Y \xrightarrow{\Sigma^V \pi} \Sigma^V K_V Y \xrightarrow{\Sigma^V \alpha_V} \Sigma^V \Omega^V \Sigma^V Y \xrightarrow{\phi_V} \Sigma^V Y.$$

The proposition gives an equivalence of infinite loop spaces between ΓCY and QY and is our preferred version of the Barratt-Quillen theorem.

[45, §14 and 15] and [46, §2 and 3] contain further discussion and various additional results about the coordinatized spectrum $B_{\infty}X$ specified by $B_iX = (\Omega^{\infty}TX)(R^i)$. Of course, the little cubes operads used in those papers could be replaced by the little convex bodies operads introduced here without any change in the results or their proofs. In particular, we have the following consistency statement, which was used in the discussion of Bott periodicity in chapter I. We give some details since the result needed there was more precise than the result proven in [45 and 46].

<u>Proposition 3.4.</u> Let X be a $\mathcal{C}'$-space. Then there is a map $\zeta : \Omega^\infty T \Omega^d X \to \Omega^d \Omega^\infty T X$ in $H\mathcal{S}$ such that the following diagram commutes in $H\mathcal{J}$.

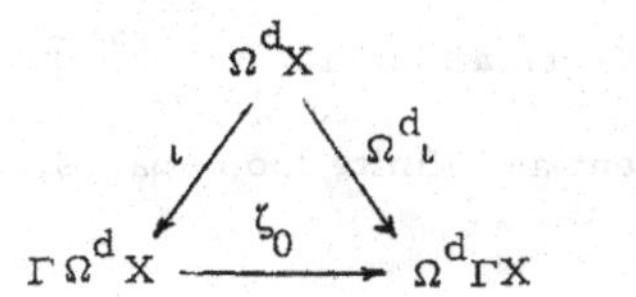

<u>Proof.</u> [46, 3.1] gives maps of $\mathcal{C}$-spaces

$$X \xleftarrow{\;\varepsilon\;} Y_d X \xrightarrow{\;\delta\;} B_d \Omega^d X$$

such that δ and $\Omega^d \varepsilon$ are weak equivalences. Think of $B_d \Omega^d X$ as the zeroth space of the coordinatized spectrum $\Omega^{-d} B_\infty \Omega^d X = \{B_{d+i} \Omega^d X\}$. Then application of the functor $\Omega^d B_\infty$ and use of Proposition 3.2 gives maps of coordinatized spectra

$$(*) \quad B_\infty \Omega^d X \cong \Omega^d \Omega^{-d} B_\infty \Omega^d X \xleftarrow{\Omega^d \tilde{\omega}} \Omega^d B_\infty B_d \Omega^d X \xleftarrow{\Omega^d B_\infty \delta} \Omega^d B_\infty Y_d X \xrightarrow{\Omega^d B_\infty \varepsilon} \Omega^d B_\infty X ,$$

the first two of which are weak equivalences. On the zeroth space level, we have the following commutative diagram of weak equivalences:

$$\begin{array}{ccccccc}
\Omega^d X & \xrightarrow{\iota} & \Omega^d B_d \Omega^d X & \xleftarrow{\Omega^d \delta} & \Omega^d Y_d X & \xrightarrow{\Omega^d \varepsilon} & \Omega^d X \\
\downarrow \iota & \overset{\Omega^d \tilde{\omega}_0}{\diagup\!\!\diagup} & \downarrow \Omega^d \iota & & \downarrow \Omega^d \iota & & \downarrow \Omega^d \iota \\
\Gamma \Omega^d X & \xleftarrow{\Omega^d \tilde{\omega}_0} & \Omega^d \Gamma B_d \Omega^d X & \xleftarrow{\Omega^d \Gamma \delta} & \Omega^d \Gamma Y_d X & \xrightarrow{\Omega^d \Gamma \varepsilon} & \Omega^d \Gamma X
\end{array}$$

Inspection of the explicit construction of the intermediate space $Y_d X$ in [45, p. 148-151 (especially the bottom diagram on p. 150)] and use of

[45,14.9] demonstrates that the composite $(\Omega^d \mathcal{E})(\Omega^d \delta)^{-1}\iota$ is equal in $H\mathcal{J}$ to the identity map of $\Omega^d X$. Now application of the functor ψ (which commutes with Ω) to the maps of (*) and use of the equivalence $\psi\phi \simeq 1$ of II.1.8 gives the required map ζ of coordinate-free spectra.

As explained in [46, 3.7 (p.75)], the previous result implies the following further consistency statement.

Proposition 3.5. Let G be a monoid in $\mathcal{C}'[\mathcal{J}]$. Then BG and the delooping $B_1 G$ are equivalent as infinite loop spaces.

§4. The two operad recognition principle

Assume given a locally contractible operad pair $(\mathcal{C}', \mathcal{G}')$, for example $(\mathcal{N}, \mathcal{G}')$ where $\mathcal{G}'$ is locally contractible or any E_∞ operad pair, and define $(\mathcal{C}, \mathcal{G})$ to be the product (partial) operad pair $(\mathcal{C}' \times \mathcal{K}_\infty, \mathcal{G}' \times \mathcal{L})$. Let $(\pi, \rho): (\mathcal{C}, \mathcal{G}) \to (\mathcal{K}_\infty, \mathcal{L})$ be the projection and regard elements of $\mathcal{G}(j)$ as linear isometries via ρ. Recall the definition, IV.1.1, of a $\mathcal{G}$-prespectrum.

Theorem 4.1. Let (X, θ, ξ) be a $(\mathcal{C}, \mathcal{G})$-space. Then $TX = T(X, \theta)$ admits a natural structure of $\mathcal{G}$-prespectrum, hence $\Omega^\infty TX$ admits a natural structure of $\mathcal{G}$-spectrum.

Proof. By IV.2.3, the second clause will follow from the first. Let V_i, $1 \leq i \leq j$, be a finite dimensional sub inner product of R^∞ and let $g \in \mathcal{G}(j)$. We must specify appropriate maps

$$(*) \quad \xi_j(g): (TX)(V_1) \wedge \ldots \wedge (TX)(V_j) \to (TX)(W), \quad \text{where } W = g(V_1 \oplus \ldots \oplus V_j).$$

These maps will be induced from composites

$$\begin{array}{c} B(\Sigma^{V_1}, C_{V_1}, X) \times \ldots \times B(\Sigma^{V_j}, C_{V_j}, X) \\ \downarrow \\ |B_*(\Sigma^{V_1}, C_{V_1}, X) \times \ldots \times B_*(\Sigma^{V_j}, C_{V_j}, X)| \\ \downarrow \\ |B_*(\Sigma^{V_1}, C_{V_1}, X) \wedge \ldots \wedge B_*(\Sigma^{V_j}, C_{V_j}, X)| \\ \downarrow |\chi_j(g)_*| \\ B(\Sigma^W, C_W, X) \end{array}$$

Here the first arrow is the natural homeomorphism of [45, 11.5], the second arrow is derived by passage to smash products in each simplicial degree, and the third arrow is the realization of a map $\chi_j(g)_*$ of simplicial spaces still to be constructed. It is apparent from Lemma 2.3 that the maps

$$\lambda : \mathcal{L}(j) \times \mathcal{K}_\infty(i_1) \times \ldots \times \mathcal{K}_\infty(i_j) \to \mathcal{K}_\infty(i_1 \cdots i_j)$$

are obtained by passage to adjoints and limits from maps

$$\lambda(\ell): \mathcal{K}_{V_1}(i_1) \times \ldots \times \mathcal{K}_{V_j}(i_j) \to \mathcal{K}_W(i_1 \cdots i_j), \quad W = \ell(V_1 \oplus \ldots \oplus V_j),$$

for $\ell \in \mathcal{L}(j)$. Therefore, in view of VI.1.8 and the product structure on $(\mathcal{C}, \mathcal{H})$, there are unique dotted arrows $\xi_j(g)$ such that the following diagrams commute (the solid arrows $\xi_j(g)$ being given by VI.1.8 and 1.9):

$$\begin{array}{ccc} C_{V_1}X \times \ldots \times C_{V_j}X & \xrightarrow{\subset} & CX \times \ldots \times CX \\ \vdots\, \xi_j(g) & & \downarrow \xi_j(g) \\ C_W X & \xrightarrow{\subset} & CX \end{array}$$

This statement holds for any $\mathcal{H}_0$-space X and in particular for CX, CCX, etc. In view of the role played in VI.1.8 by the original map

$\xi_j(g): X^j \to X$, we see that iterative application of the statement above yields unique dotted arrows $\xi_j(g)_q$, $q \geq 0$, such that the following diagrams commute:

$$\begin{array}{ccc} C^q_{V_1}X \times \ldots \times C^q_{V_j}X & \xrightarrow{C} & C^q X \times \ldots \times C^q X \\ \downarrow{\scriptstyle \xi_j(g)} & & \downarrow{\scriptstyle \xi_j(g)} \\ C^q_W X & \xrightarrow{C} & C^q X \end{array}$$

The $\xi_j(g)_q$ all pass to smash products. If we collect the smash product factors $tV_1, \ldots, tV_j$ together and apply tg to them, then we obtain from the $\xi_j(g)_q$ the further maps

$$\chi_j(g)_q : \Sigma^{V_1} C^q_{V_1} X \wedge \ldots \wedge \Sigma^{V_j} C^q_{V_j} X \longrightarrow \Sigma^W C^q_W X .$$

$B_q(\Sigma^V, C_V, X)$ is a subspace of $\Sigma^V C_V^q X$, and these maps restrict to the maps on q-simplices of the required simplicial map $\chi_j(g)_*$. The face and degeneracy operators [45, 9.6] are respected because the maps θ, μ, and η are maps of $\mathcal{L}_0$-spaces and, for the zeroth face (obtained from the $\beta_V : \Sigma^V C_V \to \Sigma^V$) because $\alpha_\infty \pi : CX \to QX$ is a map of $\mathcal{L}_0$-spaces (by Theorem 2.4) and the action of $\mathcal{L}$ on QX is induced by passage to limits from maps

$$\xi_j(g): \Omega^{V_1}\Sigma^{V_1}X \wedge \ldots \wedge \Omega^{V_j}\Sigma^{V_j}X \to \Omega^W \Sigma^W X.$$

The point is that all requisite compatibility pulls back to the level of finite dimensional inner product spaces from the compatibility statements (for K_∞ and C) codified in VI.1.6-1.9 and in section 2. It is easily verified that the maps $\xi_j(g)$ of (*) satisfy the algebraic identities specified in IV.1.1(a)-(c). Indeed, these identities are inherited from obvious identities for linear isometries and the identities given by the assertion that each $C^q X$ is a $\mathcal{L}_n$-space. Condition IV.1.1(d) holds by the continuity of all of the functors employed in the definition of the $\xi_j(g)$. An easy diagram chase shows that the compatibility condition IV.1.1(e) between the $\xi_j(g)$ and the maps σ of formula (3.5) is satisfied. Finally, condition IV.1.1(f) obviously holds if we define the map $\xi(f): X \to X$ required for formula (3.6) to be $\xi_1(g)$ for any $g \in \mathcal{H}(1)$ such that $g|V = f$.

Thus E_∞ ring spaces determine E_∞ ring spectra. We know from Theorem 3.1 how the derived additive structure (that is, the spectrum structure) is related to the given additive E_∞ space structure. We next relate the derived multiplicative structure to the given multiplicative structure.

<u>Theorem 4.2.</u> For a $(\mathcal{C}, \mathcal{H})$-space (X, θ, ξ), all of the maps $\epsilon(\theta)$, $B(\alpha_\infty \pi, 1, 1)$, γ^∞, $\tau(\eta)$, and ι specified in Theorem 3.1 are maps of $\mathcal{H}_0$-spaces, hence the first three of these maps are maps of $(\mathcal{C}, \mathcal{H})$-spaces.

<u>Proof.</u> In view of [45, 9.6 and 9.9], $\epsilon(\theta)$, $B(\alpha_\infty \pi, 1, 1)$, and $\tau(\eta)$ are geometric realizations of maps of simplicial $\mathcal{H}_0$-spaces and are therefore maps of $\mathcal{H}_0$-spaces by [45, 12.2]. The map γ^∞ is the limit over $V \subset R^\infty$ of the maps

$$\gamma^V: |B_*(\Omega^V \Sigma^V, C_V, X)| \to \Omega^V |B_*(\Sigma^V, C_V, X)|$$

(see [45, 9.7, 12.3, and 14.4]) and is easily verified to be a map of $\mathcal{H}_0$-spaces by explicit calculation (compare [45, 12.4]). It would be pointless to give the details since we know independently, by IV.1.6, that ι is a map of $\mathcal{H}_0$-spaces.

As promised, we have thus "group completed" the additive structure of an E_∞ ring space while carrying along the multiplicative structure. Again, the obvious special cases behave correctly.

Proposition 4.3. Let E be a $\mathcal{H}$-spectrum. Then $\omega: TE_0 \to E$ is a morphism of $\mathcal{H}$-prespectra, hence $\tilde{\omega}: \Omega^\infty TE_0 \to E$ is a morphism of $\mathcal{H}$-spectra.

Proof. By IV.1.1(e), we have commutative diagrams

$$\begin{array}{ccc} \Sigma^{V_1 \oplus \cdots \oplus V_j}(E_0 \wedge \cdots \wedge E_0) \cong \Sigma^{V_1} E_0 \wedge \cdots \wedge \Sigma^{V_j} E_0 & \xrightarrow{\sigma \wedge \cdots \wedge \sigma} & EV_1 \wedge \cdots \wedge EV_j \\ \downarrow \Sigma^{\rho g} \xi_j(g) & & \downarrow \xi_j(g) \\ \Sigma^W E_0 & \xrightarrow{\sigma} & EW \end{array}$$

for $g \in \mathcal{H}(j)$, $V_i \subset R^\infty$, and $W = g(V_1 \oplus \cdots \oplus V_j)$. In view of the role played by the $\xi_j(g)$ on E_0 in the definition of the $\xi_j(g)$ on TE_0, it follows readily that the diagrams

$$\begin{array}{ccc} (TE_0)(V_1) \wedge \cdots \wedge (TE_0)(V_j) & \xrightarrow{\varepsilon(\phi_{V_1}) \wedge \cdots \wedge \varepsilon(\phi_{V_j})} & EV_1 \wedge \cdots \wedge EV_j \\ \downarrow \xi_j(g) & & \downarrow \xi_j(g) \\ (TE_0)(W) & \xrightarrow{\varepsilon(\phi_W)} & EW \end{array}$$

are commutative. This proves the first part, and the second part follows by IV.1.6.

Propositions 4.3 and 3.3 imply the following result.

Corollary 4.4. For a $\mathcal{H}_0$-space Y, the composite deformation retraction

$$\Omega^\infty TCY \xrightarrow{\Omega^\infty T(\alpha_\infty \pi)} \Omega^\infty TQY \xrightarrow{\tilde{\omega}} Q_\infty Y$$

is a morphism of $\mathcal{H}$-spectra.

Indeed, even more is true. The inverse inclusion of $Q_\infty Y$ in $\Omega^\infty TCY$ and each h_t of the deformation obtained from [45, 9.9 and 11.10] by application of Ω^{∞} are also morphisms of $\mathcal{G}$-spectra. This corollary can usefully be combined with the following consequence of [46, A. 2(i)] .

Corollary 4.5. For a $\mathcal{G}_0$-space Y, the projection $\psi: CY \to C'Y$ is a map of $(\mathcal{C}, \mathcal{G})$-spaces and, if $\mathcal{C}'$ is an E_{∞} operad, the induced map $\Omega^{\infty}T\psi: \Omega^{\infty}TCY \to \Omega^{\infty}TC'Y$ of $\mathcal{G}$-spectra is a weak homotopy equivalence.

Consider, for example, the case $(\mathcal{C}', \mathcal{G}') = (\mathcal{Q}, \mathcal{Q})$. The corollaries and Proposition 3.3 imply that, for any (multiplicative) $\mathcal{Q}_0$-space Y, QY is weakly homotopy equivalent as a $(\mathcal{Q}\times\mathcal{K}_{\infty}, \mathcal{Q}\times\mathcal{L})$-space to ΓDY. When $Y = S^0$, $DY = \coprod K(\Sigma_j, 1)$ as a space. We have thus obtained a group completion of $\coprod K(\Sigma_j, 1)$ which is equivalent as an E_{∞} ring space to QS^0. This is a greatly strengthened version of the Barratt-Quillen theorem. Note that we have made no use of the monoid structures on DS^0 and our discussion applies equally well to $C'S^0 = \coprod K(\Sigma_j, 1)$ for any E_{∞} operad pair $(\mathcal{C}', \mathcal{G}')$. The force of the particular example $(\mathcal{Q}, \mathcal{Q})$ is the connection it establishes, via VI. 5. 1, between the category of finite sets and the sphere spectrum and thus, via VI. 5. 2, between algebraic K-groups and the stable homotopy groups of spheres (both with all internal structure in sight). Similarly, VI. 5. 9 and the corollaries above imply the following result.

Corollary 4.6. The E_{∞} ring spectrum $\Omega^{\infty}TB\mathcal{O}Z$ determined by the bipermutative category $\mathcal{O}Z$ is equivalent to $Q_{\infty}(RP^{\infty +})$ and, under the equivalence $\Omega^{\infty}TB\mathcal{E} \simeq Q_{\infty}S^0$, the morphism of E_{∞} ring spectra induced by the functor $e: \mathcal{E} \to \mathcal{O}Z$ coincides with the natural split injection $Q_{\infty}S^0 \to Q_{\infty}(RP^{\infty +})$.

§5. The multiplicative E_∞ structure and localization

Consider a $(\mathcal{C}, \mathcal{G})$-space (X, θ, ξ) where, as in the previous section, $(\mathcal{C}, \mathcal{G})$ is a product operad pair $(\mathcal{C}' \times \mathcal{K}_\infty, \mathcal{G}' \times \mathcal{L})$ with $\mathcal{C}'$ and $\mathcal{G}'$ locally contractible. We have a firm grasp on the $\mathcal{G}$-spectrum $\Omega^\infty T(X,\theta)$ and its relationship to X. Clearly the spectrum $\Omega^\infty T(X,\xi)$ is weakly contractible since its zeroth space is a group completion of X in which the element $0 \in \pi_0 X$ becomes invertible. Thus we must delete components of X in order to obtain interesting spectra from its multiplicative E_∞ structure.

We make a simplifying assumption. As a commutative semi-ring, $\pi_0 X$ admits a unit $e: Z^+ \rightarrow \pi_0 X$; indeed this morphism of semi-rings is obtained by application of π_0 to the unit $e: CS^0 \rightarrow X$. We assume henceforward that $e: Z^+ \rightarrow \pi_0 X$ is an inclusion (as is the case in practice). Let M be a (multiplicative) submonoid of Z^+ such that 0 is not in M and M contains at least one element other than 1. Let Z_M denote the localization of the integers at M (obtained by inverting the primes which divide elements of M). Define X_M to be the union of those components X_m of X such that $m \in M \subset \pi_0 X$ and note that X_M is a sub $\mathcal{G}$-space of X. We shall prove that the unit component of $\Gamma(X_M, \xi)$ is equivalent as an infinite loop space to the localization at M of the unit component of $\Gamma(X, \theta)$.

For an E_∞ space (Y, χ) and an element i of the group completion of $\pi_0 Y$, let $\Gamma_i(Y,\chi)$ denote the ith component of the zeroth space $\Gamma(Y,\chi)$ of $\Omega^\infty T(Y,\chi)$. Let $\iota_\oplus: X \rightarrow \Gamma(X,\theta)$ and $\iota_\otimes: X_M \rightarrow \Gamma(X_M, \xi)$ denote the group completions obtained by specialization of Theorem 3.1. We shall make one further simplifying assumption (although it could perhaps be avoided at the price of some extra work).

Definition 5.1. X is said to be convergent at M if for each prime p which does not divide any element of M there exists an eventually increasing sequence $n_i(p)$ such that $(\iota_{\oplus})_*: H_j(X_i; Z_p) \to H_j(\Gamma_i(X,\theta); Z_p)$ is an isomorphism for all $j \le n_i(p)$. Here we allow $p = 0$, when Z_p is to be interpreted as the rational numbers.

This condition seems always to be satisfied in practice.

Examples 5.2. X is convergent at M in the following cases.

(i) X is grouplike under θ, so that $\pi_0 X$ is a ring; here $\iota_{\oplus}: X_i \to \Gamma_i(X,\theta)$ is a weak homotopy equivalence.

(ii) $X = CY$ for some $\mathcal{G}_0$-space Y; here the result holds by inspection of the calculation of $H_*(CY; Z_p)$ in [26, I§5].

(iii) For p not dividing any element of M, the additive translations $X_i \to X_{i+1}$ induce isomorphisms $H_j(X_i; Z_p) \to H_j(X_{i+1}; Z_p)$ for $j \le n_i(p)$, where $\{n_i(p)\}$ is eventually increasing; here the result holds since, by [46, 3.9], $(\iota_{\oplus})_*$ induces an isomorphism

$$\varinjlim H_*(X_i; Z_p) \to \varinjlim H_*(\Gamma_i(X,\theta); Z_p) \cong H_*(\Gamma_0(X,\theta); Z_p).$$

The last example applies to $X = B\mathcal{a}$ for the interesting bipermutative categories $\mathcal{a}$ displayed in VI §5.

We shall be considering spaces obtained by application of the one operad recognition principle of Theorem 3.1 to $\mathcal{G}$-spaces, hence all spaces in sight will be $\mathcal{G} \times \mathcal{K}_\infty$-spaces (where given $\mathcal{G}$-spaces are regarded as $\mathcal{G} \times \mathcal{K}_\infty$-spaces by pullback along the projection).

We shall allow ourselves to invert weak homotopy equivalences by working in the category $H\mathcal{J}$ (see II §2).

In the case $X = QS^0$, the idea of the following result is due to Sullivan. Tornehave [76, 5.8] proved this case and also proved a somewhat weaker result in the case $X = B\mathcal{J}\mathcal{X}k_r$ [77, 3.1].

Theorem 5.3. Consider the following commutative diagram, in which all spaces are $\mathcal{J} \times \mathcal{K}_\infty$-spaces, all maps are composites of $\mathcal{J} \times \mathcal{K}_\infty$- maps and homotopy inverses of $\mathcal{J} \times \mathcal{K}_\infty$-maps, and the maps i are inclusions of components:

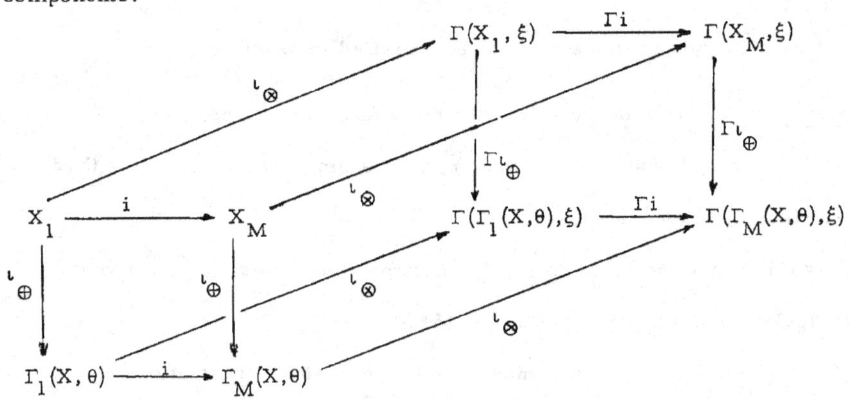

(i) If $\pi_0 X$ is a ring, then $(\Gamma i)\iota_\otimes : X_1 \to \Gamma_1(X_M, \xi)$ is a localization of X_1 at M.

(ii) If X is convergent, then $\Gamma\iota_\oplus : \Gamma(X_M, \xi) \to \Gamma(\Gamma_M(X, \theta), \xi)$ is a weak homotopy equivalence.

Therefore, if X is convergent, the composite of $\mathcal{J} \times \mathcal{K}_\infty$-maps and inverses of $\mathcal{J} \times \mathcal{K}_\infty$-maps

$$\phi = (\Gamma\iota_\oplus)^{-1}(\Gamma i)\iota_\otimes : \Gamma_1(X, \theta) \to \Gamma_1(X_M, \xi)$$

is a localization of $\Gamma_1(X, \theta)$ at M.

Proof. The last statement will follow from (ii) and from (i) applied to $\Gamma(X, \theta)$. Write the set of elements of M in order as $1, m_1, m_2, \ldots$ and define $n_i = m_1 \cdots m_i \in M$. Fix $c_n \in \mathcal{C}(n)$ and write n for $\theta_n(c_n)(1^n) \in X_n$ for any positive integer n. Consider the sequence of spaces and maps

$$(*) \qquad X_1 \xrightarrow{\tau(m_1)} X_{n_1} \xrightarrow{\tau(m_2)} X_{n_2} \rightarrow \cdots \rightarrow X_{n_{i-1}} \xrightarrow{\tau(m_i)} X_{n_i} \rightarrow \cdots$$

where $\tau(m_i)$ means multiplicative right translation by m_i; thus, for $m \in M$, $\tau(m)(x) = \xi_2(g)(x,m)$ for any fixed $g \in \mathcal{G}(2)$. By the definition of a group completion,

$$H_*(\Gamma_1(X_M, \xi); k) = \varinjlim H_*(X_n; k)$$

for any commutative ring k, where the limit is taken over

$$\tau(m)_*: H_*(X_n; k) \to H_*(X_{mn}; k), \quad m, n \in M.$$

(Compare [46, 1.2].) By cofinality, we see that $H_*(\Gamma_1(X_M, \xi); k)$ is isomorphic to $H_*(\overline{X}_M; k)$, where $\overline{X}_M$ denotes the mapping telescope of the sequence (*). Moreover, by [46, 3.9], this isomorphism can be realized naturally by a map $\overline{\iota}_\otimes : \overline{X}_M \to \Gamma_1(X_M, \xi)$ in $H\mathcal{J}$ such that the following diagram commutes in $H\mathcal{J}$ (where j is the natural inclusion):

$$\begin{array}{ccc} X_1 & \xrightarrow{\ j\ } & \overline{X}_M \\ \downarrow{\scriptstyle \iota_\otimes} & & \downarrow{\scriptstyle \overline{\iota}_\otimes} \\ \Gamma_1(X_1, \xi) & \xrightarrow{\ \Gamma i\ } & \Gamma_1(X_M, \xi) \end{array}$$

(Actually, the cited result is stated under cellular restrictions and with M free on one generator, but its proof transcribes trivially to the present context.) We prove first that $\tilde{H}_*(\overline{X}_M; Z_p) = 0$ if p divides some element of M. This will imply that multiplication by p is an isomorphism on $\tilde{H}_*(\overline{X}_M; Z)$, hence that $H_*(\overline{X}_M; Z)$ is a Z_M-module and $\Gamma_1(X_M, \xi)$ is an M-local space. $\overline{X}_M$ itself will be M-local if it is simple (or at least nilpotent). Denote the products on $H_*(X; Z_p)$ coming from θ and ξ by $x * y$ and xy and write $[n]$ for the homology class corresponding to the point $n \in X_n$. Let $x \in H_q(X; Z_p)$, $q > 0$. We claim that $x \cdot [p^t] = 0$ for t sufficiently large. Indeed, by [26, II. 1.5],

$$x \cdot [p] = x([1] * \cdots * [1]) = \sum x^{(1)} * \cdots * x^{(p)} = y^{[p]},$$

where $x \to \sum x^{(1)} \otimes \cdots \otimes x^{(p)}$ gives the iterated coproduct, y is the sum of the symmetric terms (all p $\otimes$-factors the same), and $y^{[p]}$ denotes the pth power of y under the $*$-product. Since $\deg y = q/p$, our claim follows by iteration. In the sequence (*), p divides infinitely many of the m_i and our claim therefore implies that $\tilde{H}_*(\overline{X}_M; Z_p) = 0$. To prove (i), choose points $-n \in X_{-n}$, define $\rho(n)$ to be the additive right translation $\rho(n)(x) = \theta_2(c_2)(x, n)$, and observe that the definition, VI.1.8 and VI.1.9, of a $(\mathcal{C}, \mathcal{J})$-space implies that the following ladder is homotopy commutative:

$$\begin{array}{ccccccc}
X_1 & \xrightarrow{\tau(m_1)} & X_{n_1} & \xrightarrow{\tau(m_2)} & X_{n_2} & \longrightarrow & \cdots \\
\downarrow{\scriptstyle \rho(-1)} & & \downarrow{\scriptstyle \rho(-n_1)} & & \downarrow{\scriptstyle \rho(-n_2)} & & \\
X_0 & \xrightarrow{\tau(m_1)} & X_0 & \xrightarrow{\tau(m_2)} & X_0 & \longrightarrow & \cdots
\end{array}$$

The cited definitions also imply that $\tau(m)$ is homotopic to the mth power operation $x \to \theta_m(c_m)(x^m)$. Thus the bottom arrows $\tau(m_i)$ induce multiplication by m_i on homotopy groups, hence the mapping telescope of the bottom sequence is a localization of X_0 at M. Since the vertical arrows are homotopy equivalences and $\overline{X}_M$ is simple (as a limit of simple spaces), $j: X_1 \to \overline{X}_M$ and $\overline{\iota}_{\otimes} j = (\Gamma i)\iota_{\otimes}: X_1 \to \Gamma_1(X_M, \xi)$ are also localizations at M. To prove (ii), note that the first parts of the proof apply to $\Gamma(X, \theta)$ as well as to X and consider the commutative ladder

$$\begin{array}{ccccccc}
X_1 & \xrightarrow{\tau(m_1)} & X_{n_1} & \xrightarrow{\tau(m_2)} & X_{n_2} & \longrightarrow & \cdots \\
\downarrow{\scriptstyle \iota_\oplus} & & \downarrow{\scriptstyle \iota_\oplus} & & \downarrow{\scriptstyle \iota_\oplus} & & \\
\Gamma_1(X, \theta) & \xrightarrow{\tau(m_1)} & \Gamma_{n_1}(X, \theta) & \xrightarrow{\tau(m_2)} & \Gamma_{n_2}(X, \theta) & \longrightarrow & \cdots
\end{array}$$

Since X is convergent at M, the induced map $\overline{X}_M \to \overline{\Gamma}_M(X, \theta)$ of mapping telescopes induces isomorphisms on homology with coefficients in Z_p if p does not divide any element of M. Therefore the same statement holds for

$$\Gamma\iota_{\oplus} : \Gamma_1(X_M, \xi) \rightarrow \Gamma_1(\Gamma_M(X, \theta), \xi).$$

Since these spaces are M-local, this proves (ii) on the 1-component and therefore on all components.

By application of the theorem in the situations of Corollaries 4.4 and 4.5, with $\mathcal{C}'$ an E_∞ operad, we obtain the following result. Recall from [45, 8.14] that $\pi_0 CY$ is the free commutative monoid generated by the based set $\pi_0 Y$ and that $\pi_0 QY$ is the group completion of $\pi_0 CY$. Let $C_M Y$, $C'_M Y$, and $Q_M Y$ denote the unions of the components corresponding to M in CY, C'Y and QY.

Corollary 5.4. Let Y be a $\mathcal{G}_0$-space and consider the following commutative diagram in $H\mathcal{J}$, in which all spaces are $\mathcal{G} \times \mathcal{K}_\infty$-spaces and all maps are composites of $\mathcal{G} \times \mathcal{K}_\infty$-maps and inverses of $\mathcal{G} \times \mathcal{K}_\infty$-maps:

$$\begin{array}{ccccccc}
\Gamma_1(C'Y, \theta) & \xleftarrow{\Gamma\psi} & \Gamma_1(CY, \theta) & \xrightarrow{\Gamma\alpha_\infty \pi} & \Gamma_1(QY, \theta) & \xrightarrow{\tilde{\omega}} & Q_1 Y \\
\downarrow \phi & & \downarrow \phi & & \downarrow \phi & & \downarrow (\Gamma i)\iota_{\otimes} \\
\Gamma_1(C'_M Y, \xi) & \xleftarrow{\Gamma\psi} & \Gamma_1(C_M Y, \xi) & \xrightarrow{\Gamma\alpha_\infty \pi} & \Gamma_1(Q_M Y, \xi) & = & \Gamma_1(Q_M Y, \xi)
\end{array}$$

(i) All horizontal arrows are weak homotopy equivalences.

(ii) All vertical arrows are localizations at M.

Consider the case $(\mathcal{C}', \mathcal{G}') = (\mathcal{Q}, \mathcal{Q})$ and $Y = S^0$. Here $D_M S^0 = \coprod_{m \in M} K(\Sigma_m, 1)$ and $Q_1 S^0 = SF$. Thus, as an infinite loop space, the 1-component $\Gamma_1(D_M S^0, \xi)$ of our group completion of $\coprod_{m \in M} K(\Sigma_m, 1)$ is equivalent to the localization of SF at M. This statement is a version of the main theorem of Tornehave's paper [76]. The force of the particular example $(\mathfrak{Q}, \mathfrak{Q})$ is the connection it establishes, via VI.5.1, between the category of finite sets under $\otimes$ and the theory of stable spherical fibrations.

VIII. Algebraic and topological K-theory*

We here apply the machinery of the previous two chapters to obtain E_∞ ring spectra which represent various cohomology theories of interest. The emphasis will be on the construction and analysis of approximations derived from discrete categories for spaces and spectra relevant to the J-theory diagram studied in chapter V.

In section 1, after showing that the ordinary cohomology theories with coefficients in commutative rings are represented by E_∞ ring spectra, we define higher K-groups of a permutative or bipermutative category $\mathcal{A}$ as the homotopy groups of its associated spectrum or E_∞ ring spectrum $\Omega^\infty TB\mathcal{A}$; when $\mathcal{A} = \mathcal{GL}A$ for a discrete ring A, our definition yields Quillen's higher K-groups of A [59, 61]. When A is commutative, our construction rather trivially gives the ring structure on K_*A. We have already calculated KO_*Z in VII.4.6, and, in Remarks 3.6, we shall relate this to Quillen's results about K_*Z [60]. Beyond these observations, we have no new applications to algebraic K-theory. The calculational power of infinite loop space theory lies primarily in connection with fine structure, such as homology operations (and the arguments in §4 will demonstrate how powerful this structure can be). It is not geared towards analysis of homotopy types (other than deloopings of known ones). In view of the present primitive state of calculations in algebraic K-theory, it is too early to tell how useful the rich extra structure which we shall obtain on the representing spectra for the relevant cohomology theories will turn out to be. We end section 1 with a discussion of the relationship

---

* (by J. P. May and J. Tornehave)

between representation theory and the internal structure on the zeroth spaces of spectra derived from bipermutative categories.

In section 2, we prove that the real and complex (connective) topological K-theories are represented by the E_∞ ring spectra $kO = \Omega^\infty TB\mathcal{O}$ and $kU = \Omega^\infty TB\mathcal{U}$. We then use Brauer lifting to transport Bott periodicity from kO and kU to $\Omega^\infty TB\mathcal{O}\bar{k}_q$ (q odd) and $\Omega^\infty TB\mathcal{GL}\bar{k}_q$, all completed away from q. These results imply that Brauer lifting on the completed zeroth space level is an infinite loop map, a result first proven by the second author [75], in the complex case, by different methods. We also use recent results of Adams and Priddy [8] and of Madsen, Snaith, and the second author [42], together with a representation theoretical calculation, to prove that Brauer lifting gives an infinite loop map on the multiplicative as well as on the additive infinite loop space level.

One point of these results is that they allow us to study infinite loop properties of the Adams operations, and of maps derived from them, by use of the Frobenius automorphism in section 3. In Theorem 3.2, we obtain discrete models j_p^δ for the spectra j_p introduced in V §5. These models result by completion of E_∞ ring spectra at p and, at p, the classifying space $B(SF;j_p^\delta)$ for j_p^δ-oriented spherical fibrations is $B\,\mathrm{Coker}\,J$ endowed with an infinite loop space structure. In Theorem 3.4, we use Brauer lifting to demonstrate that a large portion of the J-theory diagram, centering around $B\,\mathrm{Coker}\,J$, is a commutative diagram of infinite loop spaces and maps.

In section 4, we construct an exponential infinite loop map $\Gamma_0 B\mathcal{GL}k_r \to \Gamma_1 B\mathcal{GL}k_r$ (away from r) and prove that, with $r = r(p)$, it becomes an equivalence when localized at an odd prime p. The domain and range here are discrete models for J_p and $J_{\otimes p}$, and the constructed map factors through the unit map $e: SF \to \Gamma_1 B\mathcal{GL}k_r$ of j_p^δ. It follows that BSF splits as $BJ_p \times B\,\mathrm{Coker}\,J$ as an infinite loop space at p. These results were first proven, quite differently, by the second author [77]. The present proofs do not use Brauer lifting and illustrate the richness of structure of E_∞ ring spaces. All of the constructions we use work equally well at the prime 2, but the key calculation fails; here the orientation sequence

$$SF \xrightarrow{e} J^\delta_{\otimes 2} \xrightarrow{\tau} B(SF; j^\delta_2) \xrightarrow{q} BSF$$

(where $J^\delta_{\otimes 2}$ is the 1-component of the zeroth space of j^δ_2) may be regarded as a codification of how the infinite loop space SF is built up from $B\,\mathrm{Coker}\,J$ and a discrete model for $J_{\otimes 2}$.

We agree to replace any space not of the homotopy type of a CW-complex by a weakly equivalent CW-complex, without change of notation (so as to allow the construction of inverse maps to weak equivalences without further verbiage).

§1. Examples; algebraic K-theory

Let A be a commutative topological semi-ring or, equivalently, an $(\mathcal{N},\mathcal{N})$-space (see VI. 2. 4). By VII. 4. 1 and VII. 2. 4, $\Omega^\infty TA$ is an E_∞ ring spectrum (in fact, since $\mathcal{N} \times \mathcal{L} = \mathcal{L}$, an $\mathcal{L}$-spectrum) and its zeroth space ΓA is an E_∞ ring space. By VII. 4. 2, $\iota : A \to \Gamma A$ is a group completion of the additive structure of A which is compatible with its multiplicative structure. Of course, ΓA is not a ring. The original precise structure on A has been weakened to structure precise only up to all higher coherence homotopies.

Now let A be discrete. Then ΓA is homologically discrete, in the sense that $H_i \Gamma A = 0$ for $i > 0$, and $\iota_* : A = \pi_0 A \to \pi_0 \Gamma A$ is the completion of A to a ring. If A is already a ring, then $\iota : A \to \Gamma A$ is a homotopy equivalence and the E_∞ ring spectrum $\Omega^\infty TA$ is thus an Eilenberg-Mac Lane spectrum $HA = \mathcal{K}(A, 0)$. Therefore any ordinary cohomology theory with coefficients in a commutative ring is represented by an E_∞ ring spectrum.

Less trivial examples arise from categories associated to commutative rings. We proceed from the general to the particular.

If $(\mathcal{A}, \oplus, 0, c)$ is a permutative category, then $B\mathcal{A}$ is a $\mathcal{Q}$-space, by VI. 4. 2, and VII. 3.1 gives an infinite loop space $\Gamma B\mathcal{A}$ and a map $\iota : B\mathcal{A} \to \Gamma B\mathcal{A}$ which is a group completion. We define the algebraic K-groups of $\mathcal{A}$ by

$$K_i\mathcal{A} = \pi_i(\Gamma B\mathcal{A}, 0) \quad \text{for } i \geq 0.$$

If $(\mathcal{A}, \oplus, 0, c, \otimes, 1, \tilde{c})$ is a bipermutative category, then $B\mathcal{A}$ is a $(\mathcal{Q}, \mathcal{Q})$-space, by VI. 4. 4, and VII. 2. 4 and VII. 4. 2 give that $\Gamma B\mathcal{A}$ is a $(\mathcal{Q} \times \mathcal{K}_\infty, \mathcal{Q} \times \mathcal{L})$-space and that ι is compatible with the multiplicative as well as the additive structure. Moreover, VI. 2. 5 gives that $K_*\mathcal{A}$ is a commutative (in the graded sense) and associative ring with unit. Additive right translation by one defines a homotopy equivalence $\rho(1)$ from the zero component $\Gamma_0 B\mathcal{A}$ to the 1-component $\Gamma_1 B\mathcal{A}$. Since $(\Gamma_0 B\mathcal{A}, \theta)$ and $(\Gamma_1 B\mathcal{A}, \xi)$ are E_∞ spaces, we therefore have two 0-connected spectra, one coming from $\oplus$ and the other from $\otimes$, both of which have the higher K-groups of $\mathcal{A}$ as homotopy groups. These spectra will generally be very different, but Theorem 4.1 below will show that in certain interesting cases they do become equivalent when localized at an appropriate prime.

Now let the permutative category $\mathcal{A}$ be of the form specified in VI. 5. 8, so that $\mathcal{A}$ can be thought of as a disjoint union of topological groups $G(n)$ for $n \geq 0$. Then $B\mathcal{A} = \coprod BG(n)$. Define BG to be the limit of the translations $\rho(1): BG(n) \to BG(n+1)$. As explained in [46, 3.9], there is a well-defined natural homotopy class

$$\bar{\iota}: BG \to \Gamma_0 B\mathcal{A}$$

such that the restriction of $\bar{\iota}$ to $BG(n)$ is homotopic to the composite of $\iota: BG(n) \to \Gamma_n B\mathcal{A}$ and the translation $\Gamma_n B\mathcal{A} \to \Gamma_0 B\mathcal{A}$; the fact that $\iota: B\mathcal{A} \to \Gamma B\mathcal{A}$ is a group completion implies that $\bar{\iota}$ induces an isomorphism on homology (with any coefficients). Therefore ΓBG is homologically equivalent to $BG \times Z$.

To relate the constructions above to Quillen's algebraic K-theory, we must review some of his results and definitions [59,61, 62]. Recall that a group is said to be perfect if it is equal to its commutator subgroup. Let X be a connected CW-complex and let N be a perfect normal subgroup of $\pi_1 X$. Then there is a map $f: X \to X^+$, unique up to homotopy, such that the kernel of $\pi_1 f$ is N and f induces an isomorphism on homology with any coefficients (see Wagoner [79]). If N is the commutator subgroup of $\pi_1 X$, then X^+ is a simple space. If Y is a connected space such that $\pi_1 Y$ contains no non-trivial perfect subgroup, then $f^*: [X^+, Y] \to [X, Y]$ is an isomorphism.

In this connection, we record the following useful triviality.

<u>Lemma 1.1.</u> Let $f: X \to X'$ be a map of connected spaces which induces an isomorphism on integral homology. Then

$$f^*: [X', \Omega Z] \to [X, \Omega Z]$$

is an isomorphism for any space Z.

<u>Proof.</u> $\Sigma f: \Sigma X \to \Sigma X'$ is an equivalence.

Let A be a discrete ring (associative with unit). The commutator subgroup EA of GLA is perfect, and Quillen defined

$$K_iA = \pi_i(BGLA^+) \quad \text{for } i \geq 1.$$

Consider the permutative category $\mathcal{GL}A$ of VI. 5.2. By the universal property of f (or the lemma), $\bar{\iota}: BGLA \to \Gamma_0 B\mathcal{GL}A$ induces a map $\tilde{\iota}: BGLA^+ \to \Gamma_0 B\mathcal{GL}A$ such that $\tilde{\iota} \circ f$ is homotopic to $\bar{\iota}$. Since $\bar{\iota}$ and f are homology isomorphisms, so is $\tilde{\iota}$. Since $BGLA^+$ and $\Gamma_0 B\mathcal{GL}A$ are simple, $\tilde{\iota}$ is therefore a homotopy equivalence. Thus

$$K_iA = K_i\mathcal{GL}A \quad \text{for } i \geq 1.$$

Now let A be commutative. Then $\mathcal{GL}A$ is bipermutative and $K_*\mathcal{GL}A$ is thus a ring. Here $K_0\mathcal{GL}A = Z$. If instead of $\mathcal{GL}A$ we use a bipermutative category $\mathcal{P}A$ of finitely generated projective modules (as exists by VI. 3. 5), then we obtain a commutative graded ring $K_*\mathcal{P}A$ such that $K_i\mathcal{P}A = K_iA$ for all $i \geq 0$ (by [46, p.85]).

Alternative constructions of spectra having the K_iA as homotopy groups can be obtained by use of the black boxes of Boardman and Vogt, Segal, Anderson, the second author, and Barratt and Eccles [20,68,10,76, 16]. It seems likely that all of these constructions yield spectra equivalent to ours, but a proof would be tedious and unrewarding. Gersten and Wagoner [30 and 79] have constructed a spectrum having the K_iA as homotopy groups by means of ring theoretic arguments within algebraic K-theory. The relationship between their spectrum and ours will be determined in chapter IX.

The discussion of sections 2 and 3 below suggests that the algebraic K-theory of discrete commutative rings can be thought of as analogous to complex K-theory and that the appropriate analog of real K-theory can be defined by

$$KO_iA = K_i\mathcal{O}A \quad \text{for } i \geq 1,$$

where $\mathcal{O}A$ is the bipermutative cagegory of VI. 5.3. The ring homomorphism

$$K_*\mathcal{O}A \to K_*\mathcal{H}\mathcal{L}A$$

induced by the inclusion of $\mathcal{O}A$ in $\mathcal{H}\mathcal{L}A$ can be thought of as analogous to complexification. This idea is presumably not new: it can be viewed as the starting point for Karoubi's treatment of Hermitian K-theory [34]. (However, by VII. 4. 6, KO_*Z is not very interesting.)

The following immediate consequence of VI. 5. 8, VII. 3. 1, and VII. 4. 2 plays a key role in many of the topological applications. It reduces to group theory the analysis of the action maps θ and ξ on the E_∞ ring space $\Gamma B\mathcal{A}$ derived from a bipermutative category $\mathcal{A} = \coprod G(n)$. Let

$$c_p: \Sigma_p \int G(n) \to G(pn) \quad \text{and} \quad \tilde{c}_p: \Sigma_p \int G(n) \to G(p^n)$$

be the homomorphisms of groups specified by

$$c_p(\sigma; g_1, \ldots, g_p) = (g_{\sigma^{-1}(1)} \oplus \ldots \oplus g_{\sigma^{-1}(p)})\sigma(n, \ldots, n)$$

and

$$\tilde{c}_p(\sigma; g_1, \ldots, g_p) = (g_{\sigma^{-1}(1)} \times \ldots \times g_{\sigma^{-1}(p)})\sigma\langle n, \ldots, n\rangle.$$

(See VI. 1. 1 and 1. 4 for notations.) Let ψ denote both projections $\mathcal{Q} \times \mathcal{K}_\infty \to \mathcal{Q}$ and $\mathcal{Q} \times \mathcal{L} \to \mathcal{Q}$.

<u>Proposition 1. 2.</u> For any permutative category $\mathcal{A}$ of the form $\coprod G(n)$, the following diagram is homotopy commutative:

$$\begin{array}{ccc}
[\mathcal{Q}(p) \times \mathcal{K}_\infty(p)] \times_{\Sigma_p} BG(n)^p \xrightarrow{\psi \times 1} \mathcal{Q}(p) \times_{\Sigma_p} BG(n)^p \cong B(\Sigma_p \int G(n)) & \xrightarrow{Bc_p} & BG(pn) \\
\downarrow 1 \times \iota^p & & \downarrow \iota \\
[\mathcal{Q}(p) \times \mathcal{K}_\infty(p)] \times_{\Sigma_p} (\Gamma_n B\mathcal{A})^p & \xrightarrow{\theta_p} & \Gamma_{pn} B\mathcal{A}
\end{array}$$

If $\mathcal{A}$ is bipermutative, then the following diagram is commutative:

$$\begin{array}{ccc}
[\mathcal{Q}(p)\times\mathcal{L}(p)]\times_{\Sigma_p} BG(n)^p \xrightarrow{\psi\times 1} \mathcal{Q}(p)\times_{\Sigma_p} BG(n)^p \cong B(\Sigma_p\int G(n)) & \xrightarrow{B\check{c}_p} & BG(n^p) \\
\downarrow 1\times\iota^p & & \downarrow \iota \\
[\mathcal{Q}(p)\times\mathcal{L}(p)]\times_{\Sigma_p}(\Gamma_n B\mathcal{A})^p & \xrightarrow{\xi_p} & \Gamma_{n^p}B\mathcal{A}
\end{array}$$

Let $\mathcal{A} = \mathcal{GL}A$ for a commutative topological ring A and let G be a compact topological group. For a representation $\rho: G \to GL(n, A)$ and a subgroup π of Σ_p, define the additive and multiplicative wreath product representations $\pi\int\rho$ and $\pi\int_{\otimes}\rho$ to be the composites

$$\pi\int G \xrightarrow{1\int\rho} \pi\int GL(n, A) \xrightarrow{c_p} GL(pn, A)$$

and

$$\pi\int G \xrightarrow{1\int\rho} \pi\int GL(n, A) \xrightarrow{\tilde{c}_p} GL(n^p, A)$$

The proposition reduces analysis of θ_p and ξ_p on $\Gamma B\mathcal{GL}A$ to analysis of these wreath products. Pragmatically, however, there is an essential difference. The operation $\pi\int\rho$ is additive in ρ, hence passes to representation rings, and is trivially seen to satisfy the character formula

$$\chi(\pi\int\rho)(\sigma; g_1, \ldots, g_p) = \sum_{\sigma(i) = i} \chi(\rho)(g_i) \;.$$

In contrast, $\pi\int_{\otimes}\rho$ is multiplicative but not additive in ρ,[1] and there is no general formula for the calculation of $\chi(\pi\int_{\otimes}\rho)$ in terms of $\chi(\rho)$.

Of course, E_∞ maps are structure-preserving before passage to homotopy, whereas representation theoretic techniques apply only after passage to homotopy. This suggests use of the following (not quite standard) definition.

---

[1] With $k = p$ and $j_r = 2$, the diagram of VI. 4.3 implies the formula

$$\pi\int_{\otimes}(\rho+\sigma) = \pi\int_{\otimes}\rho + \pi\int_{\otimes}\sigma + \sum_{i=1}^{p-1}\frac{1}{p}(i, p-i)\pi\int\rho^i\sigma^{p-i} \;.$$

Definition 1.3. Let p be a prime, let π be the cyclic group of order p embedded as usual in Σ_p, and let W be any contractible space on which π acts freely (for example $\mathcal{C}(p)$ for any E_∞ operad $\mathcal{C}$). An H^p_∞-space (X, θ) is a (homotopy associative) H-space X together with a map $\theta: W \times_\pi X^p \to X$ such that for each $w \in W$ the restriction of θ to $X^p \cong w \times X^p$ is homotopic to the p-fold iterate of the product on X. An H^p_∞-map $f: (X, \theta) \to (X', \theta')$ is an H-map $f: X \to X'$ such that the following diagram is homotopy commutative:

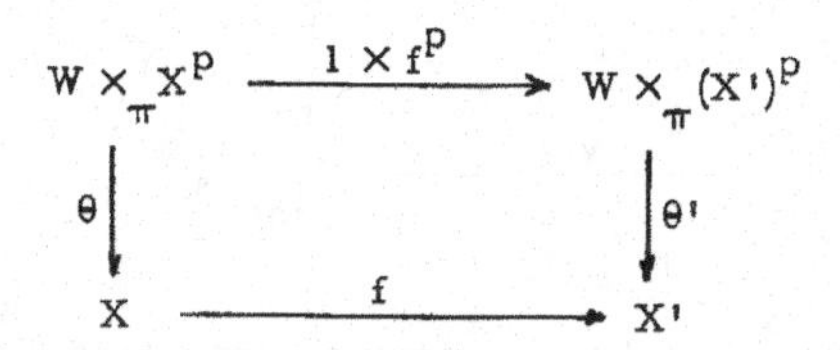

Clearly an E_∞ map, and in particular an infinite loop map, is an H^p_∞-map. Mod p homology operations are defined in terms of θ_* [26, I§1] and are thus preserved by H^p_∞-maps. If X and X' are infinite loop spaces derived from permutative categories (of the usual form) and if appropriate $\varprojlim^1$ terms vanish, then Proposition 1.2 reduces the determination of whether or not an H-map $f: X \to X'$ is an H^p_∞-map to representation theory. The following remarks give the details of this reduction.

Remarks 1.4. Let Y be an infinite loop space with induced H^p_∞-structure $\xi: W \times_\pi Y^p \to Y$. (We use the letter ξ since we choose to think of Y as multiplicative, that being appropriate to our applications of these remarks in the next section.) Fix points m and m^{-1} in the components Y_m and $Y_{m^{-1}}$ of Y. Since the product $\#$ and inverse map χ on Y are infinite loop maps and thus H^p_∞-maps, the following diagrams are homotopy commutative; they show how ξ on $W \times_\pi Y_1^p$ relates to ξ on $W \times_\pi Y_m^p$:

$$\begin{array}{ccccc} W\times_\pi Y_1^p & \xrightarrow{1\times\tau(m)^p} & W\times_\pi Y_m^p & \xrightarrow{(1,\omega_m^{-1})} & (W\times_\pi Y_m^p)\times(W\times_\pi\{m^{-1}\}^p)\,, \\ \downarrow{\scriptstyle \xi} & & & & \downarrow{\scriptstyle \xi\times\xi} \\ Y_1 & & \xleftarrow{\#} & & Y_{m^p}\times Y_{m^{-p}} \end{array}$$

where $\tau(m)(y) = y \# m$ and $\omega_m(w, y_1, \ldots, y_p) = (w, m, \ldots, m)$, and

$$\begin{array}{ccc} W\times_\pi Y_{m^{-1}}^p & \xrightarrow{1\times\chi^p} & W\times_\pi Y_m^p \\ \downarrow{\scriptstyle \xi} & & \downarrow{\scriptstyle \xi} \\ Y_{m^{-p}} & \xleftarrow{\chi} & Y_{m^p} \end{array}$$

These diagrams combine to give the homotopy commutative diagram

$$\begin{array}{ccccc} W\times_\pi Y_1^p & \xrightarrow{1\times\tau(m)^p} & W\times_\pi Y_m^p & \xrightarrow{(1,\omega_m)} & (W\times_\pi Y_m^p)\times(W\times_\pi\{m\}^p) \\ \downarrow{\scriptstyle \xi} & & & & \downarrow{\scriptstyle \xi\times\xi} \\ Y_1 & \xleftarrow{\#} & Y_{m^p}\times Y_{m^{-p}} & \xleftarrow{1\times\chi} & Y_{m^p}\times Y_{m^p} \end{array}$$

Let Z be another infinite loop space and suppose that $\pi_0 Y = \pi_0 Z$. Let $f\colon Y_1 \to Z_1$ be an H-map. Since $Y \simeq Y_1 \times \pi_0 Y$ and $Z \simeq Z_1 \times \pi_0 Z$ as H-spaces [26, I.4.6], f is the 1-component of an H-map $f\colon Y \to Z$ with $\pi_0 f = 1$. Assume that $Y = \Gamma(X, \xi)$, where (X, ξ) is an E_∞ space with $\pi_0 X$ a multiplicative submonoid of Z^+, and let $\iota\colon X \to Y$ be the natural E_∞ map. Write m for consistently chosen basepoints in X_m, Y_m, and Z_m. All parts of the following diagram commute, by the facts above, except for the front and lower back right rectangles:

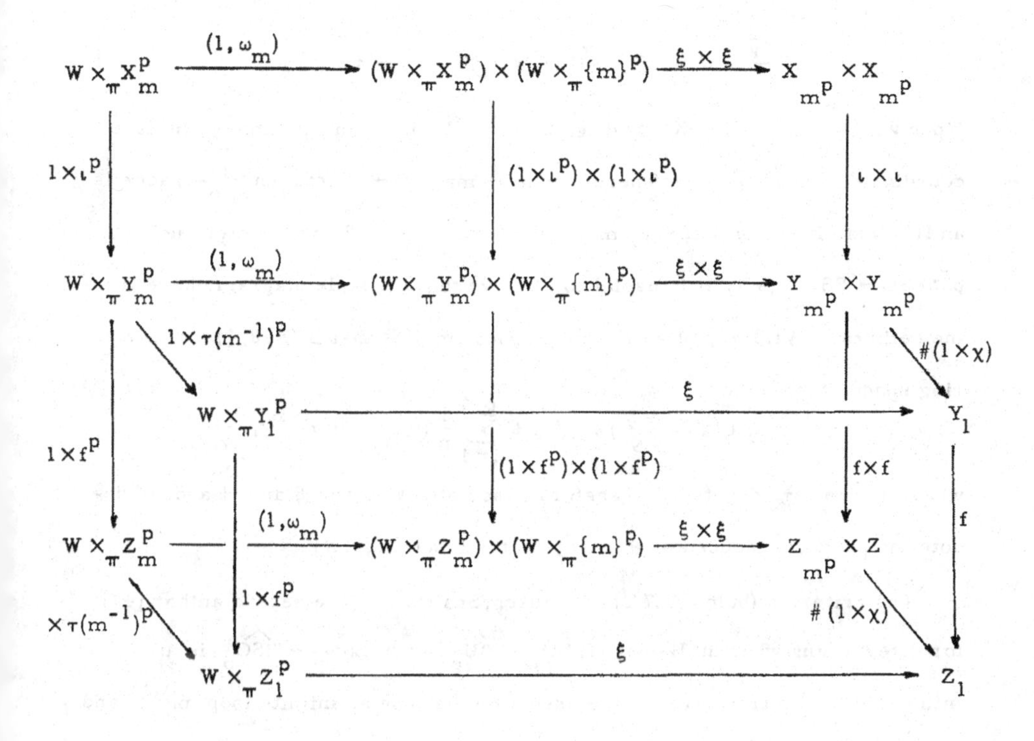

By [46, 3.9], if $\overline{X}$ is the mapping telescope of (a cofinal subsequence of) the X_m, then there is a homology isomorphism $\overline{\iota} : \overline{X} \to Y_1$ whose restriction to X_m is homotopic to $X_m \xrightarrow{\iota} Y_m \xrightarrow{\tau(m^{-1})} Y_1$. $1 \times \overline{\iota}^p : W \times_\pi \overline{X}^p \to W \times_\pi Y_1^p$ and the natural map $\mathrm{Tel}(W \times_\pi X_m^p) \to W \times_\pi \overline{X}^p$ are homology isomorphisms, hence, by Lemma 1.1 and the diagram, we conclude that if $\varprojlim^1 [W \times_\pi X_m^p, \Omega Z_1] = 0$ and if $f\iota : X \to Z$ is an H_∞^p-map, then $f : Y_1 \to Z_1$ is an H_∞^p-map. When the X_m are classifying spaces of groups and Z_m is $BO \times \{m\}$ or $BU \times \{m\}$, Proposition 1.2 reduces the verification that $f\iota$ is an H_∞^p-map to representation theory.

The following remarks recast the notions of H_∞^p-space and H_∞^p-map.

<u>Remarks 1.5.</u> Let (X, θ) be an H_∞^p-space. Let $\nu : E \to B$ be a principal π-covering classified by $\mu : B \to W/\pi$, let $\tilde{\mu} : E \to W$ cover μ, and define $\overline{\nu} : B \to W \times_\pi E^p$ by $\overline{\nu}(b) = (\tilde{\mu}e, \alpha e, \ldots, \alpha^p e)$, where $\nu e = b$ and α generates π. For $g \in [E^+, X]$, define the transfer $\tau(g) \in [B^+, X]$ to be the composite

$$B \xrightarrow{\bar{\nu}} W \times_{\pi} E^p \xrightarrow{1 \times g^p} W \times_{\pi} X^p \xrightarrow{\theta} X .$$

When ν is $W \times X^p \to W \times_{\pi} X^p$ and $g: W \times X^p \to X$ is the projection on the last coordinate, $(1 \times g^p)\bar{\nu} = 1$. Therefore an H-map $X \to Y$ between H_{∞}^p-spaces is an H_{∞}^p-map if and only if it commutes with transfer. For examples such as $\rho^r: BSO \to BSO_{\otimes}$ [42], it is useful to observe that a simple diagram chase from the definition, VI.1.10, of an E_{∞} ring space implies that if (X, θ, ξ) is an E_{∞} ring space, then

$$\tau_{\otimes}(f+g) = \tau_{\otimes}(f) + \tau_{\otimes}(g) + \sum_{i=1}^{p-1} \frac{1}{p}(i, p-i)\tau(f^i g^{p-i}) ,$$

where τ and $\tau_{\otimes}$ denote the transfers associated to θ and ξ and $+$ and $\cdot$ denote the products induced by θ and ξ on the functor $[?, X]$.

The criterion (below V.7.2) of Madsen, Snaith, and the second author [42] for determining when an H-map $f: \widehat{BU}_p \to \widehat{BU}_p$ or $f: \widehat{BSO}_p \to \widehat{BSO}_p$ is an infinite loop map translates to the assertion that f is an infinite loop map if and only if it is an H_{∞}^p-map. The Adams-Priddy theorem V.4.2, together with II.2.13 and 2.14, yields the following very useful consequence.

Theorem 1.6. Let X and Y be infinite loop spaces of the homotopy type of BSO or of BSU localized or completed at a set of primes T. Then an H-map $X \to Y$ is an infinite loop map if and only if its completion at p is an H_{∞}^p-map for all primes $p \in T$.

§2. Bott periodicity and Brauer lifting

Write $\mathscr{G}$ for either of the bipermutative categories $\mathscr{O}$ or $\mathscr{U}$ specified in VI.5.4 and write G for either O or U. Define $kG = \Omega^{\infty} TB\mathscr{G}$. The homology isomorphism $\bar{\iota}: BG \to \Gamma_0 B\mathscr{G}$ is a homotopy equivalence, hence the zeroth space $\Gamma B\mathscr{G}$ of kG is equivalent to $BG \times Z$. Since $\iota: \coprod BG(n) \to \Gamma BG$ is a map of H-spaces for both $\oplus$ and $\otimes$, it follows that ΓBG represents the ring-valued functor KGX on finite dimensional CW-complexes X. The external tensor product $KGX \otimes KGY \to KG(X \times Y)$ is defined on maps $f: X^+ \to \Gamma B\mathcal{G}$ and $g: Y^+ \to \Gamma B\mathcal{G}$ (where the plus notation again denotes addition of a disjoint basepoint) as the composite

$$(X \times Y)^+ = X^+ \wedge Y^+ \xrightarrow{f \wedge g} \Gamma B\mathcal{G} \wedge \Gamma B\mathcal{G} \xrightarrow{\otimes} \Gamma B\mathcal{G} .$$

Let $d = 8$ or $d = 2$ for $G = O$ or $G = U$. One formulation of Bott periodicity asserts that

$$\otimes : KGX \otimes KGS^d \to KG(X \times S^d)$$

is an isomorphism or, equivalently, that tensoring with a generator $[b] \in \pi_d BG = \pi_d \Gamma_0 B\mathcal{G}$ defines an isomorphism

$$\widetilde{K}GX \to \widetilde{K}G(\Sigma^d X).$$

Bott [21] deduced the latter isomorphism by verifying that the adjoint $BG \to \Omega_0^d BG$ of the composite

$$BG \wedge S^d \xrightarrow{1 \wedge b} BG \wedge BG \xrightarrow{\otimes} BG$$

is homotopic to the iterated Bott map discussed in I§1. Under the equivalence $\bar{\iota} : BG \to \Gamma_0 BG$, this adjoint corresponds to the restriction to 0-components of zeroth spaces of the adjoint $\beta : kG \to \Omega^d kG$ of the composite map in $H\mathcal{J}$

$$kG \wedge S^d \xrightarrow{1 \wedge b} kG \wedge \Gamma_0 B\mathcal{G} \xrightarrow{\phi} kG$$

given by II. 2. 9. In view of II. 3. 10, we conclude that $\beta_0 : \Gamma B\mathcal{G} \to \Omega^d \Gamma B\mathcal{G}$ agrees under the equivalence $\Gamma B\mathcal{G} \simeq BG \times Z$ with the Bott map $BG \times Z \to \Omega^d(BG \times Z)$.

By II. 3. 2, II. 3. 9, and II. 3. 14, it follows immediately that kO and kU are isomorphic in $H\mathcal{J}$ to the connective ring spectra obtained from the periodic Bott spectra by killing their homotopy groups in negative degrees. We have thus proven the following result.

Theorem 2.1. kO and kU represent real and complex connective K-theory (as ring-valued cohomology theories).

In particular, the diagrams of Proposition 1.2 now reduce the analysis of homology operations on BO and BU to representation theory. Application of the additive diagram to BO and BU was first justified by Boardman [unpublished] and has been exploited by Priddy [54] in mod 2 homology and by Snaith [70] in K-theory.

Remark 2.2. As proven by Bott [21], real periodicity factors as the composite of the two natural isomorphisms

$$\widetilde{K}OX \otimes \widetilde{K}SpS^4 \to \widetilde{K}Sp(\Sigma^4 X) \quad \text{and} \quad \widetilde{K}SpY \otimes \widetilde{K}SpS^4 \to \widetilde{K}O(\Sigma^4 Y).$$

A full understanding of these transformations in our context would seem to require a theory of E_∞ module spectra over E_∞ ring spectra.

We now turn to Brauer lifting. Fix a prime q and let $k = \overline{k}_q$ be an algebraic closure of the field of q elements. Let k_r denote the field with $r = q^a$ elements contained in k, so that $k = \lim\limits_{\rightarrow} k_r$. Fix an embedding $\mu: k^* \to \mathbb{C}^*$ of multiplicative groups. Recall from Green [31, Theorem 1] that if $\rho: G \to GL(n,k)$ is a representation of a finite group G and if $\rho(g)$ has roots $\xi_i(g)$, then the complex-valued function

$$\chi_\rho(g) = \sum_{i=1}^{n} \mu\xi_i(g)$$

is the character of a unique (virtual) representation $\lambda(\rho) \in RG$. Quillen [58, p. 79] proved that if q is odd and ρ takes values in $O(n,k)$, then χ_ρ is the character of a (necessarily unique) real representation $\lambda(\rho) \in ROG$. If $\pi: H \to GL(m,k)$ is a representation of another finite group H, then

$$\chi_{\pi\oplus\rho}(h,g) = \chi_\pi(h) + \chi_\rho(g) \quad \text{and} \quad \chi_{\pi\otimes\rho}(h,g) = \chi_\pi(h)\chi_\rho(g).$$

Therefore, if $R_k G$ denotes the representation ring of G over k, the following diagrams are commutative:

$$\begin{array}{ccc} R_kH \times R_kG & \xrightarrow{\oplus} & R_k(H\times G) \\ \downarrow{\scriptstyle \lambda\times\lambda} & & \downarrow{\scriptstyle \lambda} \\ RH \times RG & \xrightarrow{\oplus} & R(H\times G) \end{array} \quad \text{and} \quad \begin{array}{ccc} R_kH \times R_kG & \xrightarrow{\otimes} & R_k(H\times G) \\ \downarrow{\scriptstyle \lambda\times\lambda} & & \downarrow{\scriptstyle \lambda} \\ RH \times RG & \xrightarrow{\otimes} & R(H\times G) \end{array} \qquad \text{(A)}$$

By Adams' formula [1, 4.1(vi)], the following diagram also commutes, where, for $r = q^a$, ϕ^r denotes the iterated Frobenius automorphism (VI. 5.5):

$$\begin{array}{ccc} R_kG & \xrightarrow{\phi^r} & R_kG \\ \downarrow{\scriptstyle \lambda} & & \downarrow{\scriptstyle \lambda} \\ RG & \xrightarrow{\psi^r} & RG \end{array} \qquad \text{(B)}$$

Thus $\psi^r\lambda(\rho) = \lambda(\rho)$ if $\rho: G \to GL(n,k)$ factors through $GL(n,k_r)$. The analogs of the diagrams above also commute in the real case and relate the orthogonal representation ring RO_kG to ROG.

Of course, passage to classifying maps and then to Grothendieck groups gives ring homomorphisms $RG \to KU(BG)$ and $ROG \to KO(BG)$, and these become isomorphisms when the left sides are completed with respect to the IG-adic topology [14]. Moreover, by [14, 4.2 and 7.1 (and p. 13, 17)], $KU^{-1}(BG) = 0$ and $KO^{-1}(BG)$ is a finite dimensional vector space over Z_2. Let $\lambda(n,r): BGL(n;k_r) \to BU$ represent the element of $KU(BGL(n;k_r))$ obtained by application of λ to the difference of the inclusion of $GL(n,k_r)$ in $GL(n,k)$ and the trivial representation of degree n. Since λ is natural in G and additive, the maps $\lambda(n,r)$ are compatible (up to homotopy) as n and r increase. Since the relevant $\varprojlim^1$ terms vanish, there result unique homotopy classes

(*) $\lambda : BGL\overline{k}_q \to BU$ and, if q is odd, $\lambda : BO\overline{k}_q \to BO$

compatible with the $\lambda(n, r)$. The main step in Quillen's proof of the Adams conjecture was the following result [58, 1.6].

Theorem 2.3. The maps λ of (*) induce isomorphisms on cohomology with coefficients in Z_p for each prime $p \neq q$.

As we have explained in section 1, the group completion property of the recognition principle VII.3.1 gives homology isomorphisms

$$\overline{\iota} : BGLk \to \Gamma_0 B\mathcal{GL}k \quad \text{and} \quad \overline{\iota} : BOk \to \Gamma_0 B\mathcal{O}k .$$

Invoking Lemma 1.1, we define

(**) $\tilde{\lambda} : \Gamma_0 B\mathcal{GL}\overline{k}_q \to BU$ and, if q is odd, $\tilde{\lambda} : \Gamma_0 B\mathcal{O}\overline{k}_q \to BO$

to be the unique homotopy classes such that $\tilde{\lambda} \circ \overline{\iota} \simeq \lambda$. (Of course, we could also invoke the properties of the plus construction, but its use would add nothing to the discussion.)

We shall need the following observation.

Lemma 2.4. The following diagrams are homotopy commutative:

$$\begin{array}{ccc} \Gamma_0 B\mathcal{GL}k \times \Gamma_0 B\mathcal{GL}k & \xrightarrow{\oplus} & \Gamma_0 B\mathcal{GL}k \\ \downarrow{\scriptstyle \tilde{\lambda}\times\tilde{\lambda}} & & \downarrow{\scriptstyle \tilde{\lambda}} \\ BU \times BU & \xrightarrow{\oplus} & BU \end{array} \quad \text{and} \quad \begin{array}{ccc} \Gamma_0 B\mathcal{GL}k \wedge \Gamma_0 B\mathcal{GL}k & \xrightarrow{\otimes} & \Gamma_0 B\mathcal{GL}k, \\ \downarrow{\scriptstyle \tilde{\lambda}\wedge\tilde{\lambda}} & & \downarrow{\scriptstyle \tilde{\lambda}} \\ BU \wedge BU & \xrightarrow{\otimes} & BU \end{array}$$

and similarly with $\Gamma_0 B\mathcal{GL}k$ and BU replaced by $\Gamma_0 B\mathcal{O}k$ and BO.

Proof. Since $\widetilde{KU}(X \wedge Y) \to \widetilde{KU}(X \times Y)$ is a monomorphism for any X and Y, it suffices to consider the second diagram with smash products replaced by Cartesian products. For both diagrams, it suffices to prove commutativity after composition with $\overline{\iota} \times \overline{\iota}$ and thus, since the relevant $\varprojlim^1$ terms vanish, after further composition with the inclusions of $BGL(m, k_r) \times BGL(n, k_s)$. Now the conclusion is immediate from the diagrams (A).

Similarly, diagram (B) implies the following result:

Lemma 2.5. The following diagrams are homotopy commutative:

$$\begin{array}{ccc} \Gamma_0 B\mathcal{GL}k & \xrightarrow{\phi^r} & \Gamma_0 B\mathcal{GL}k \\ \downarrow \tilde{\lambda} & & \downarrow \tilde{\lambda} \\ BU & \xrightarrow{\psi^r} & BU \end{array} \quad \text{and} \quad \begin{array}{ccc} \Gamma_0 B\mathcal{O}k & \xrightarrow{\phi^r} & \Gamma_0 B\mathcal{O}k \\ \downarrow \tilde{\lambda} & & \downarrow \tilde{\lambda} \\ BO & \xrightarrow{\psi^r} & BO \end{array}$$

At this point, it will be convenient to introduce a generic (and abusive) simplification of notation, to be used throughout the rest of the book. We shall write Y^δ for specified "discrete models" for topologically significant spaces or spectra Y. In each case, Y^δ will be derived from the classifying spaces of discrete categories by means of suitable topological constructions. In particular, we have the following notations.

Definition 2.6. Define BU^δ and BO^δ to be the completions away from q of the spaces $\Gamma_0 B\mathcal{GL}\bar{k}_q$ and (with q odd) $\Gamma_0 B\mathcal{O}\bar{k}_q$. Revert to the convention that $G = O$ or $G = U$ and define $\hat{\lambda} : BG^\delta \to \widehat{BG}[1/q]$ to be the completion away from q of the map $\tilde{\lambda}$ of (**). Define kO^δ and kU^δ to be the completions away from q of the E_∞ ring spectra $\Omega^\infty TB\mathcal{O}\bar{k}_q$ and $\Omega^\infty TB\mathcal{GL}\bar{k}_q$. Then BG^δ is the 0-component of the zeroth space of kG^δ.

The following result is an immediate consequence of Theorem 2.3 and the homological characterization of completions (see II §2).

Corollary 2.7. $\hat{\lambda} : BG^\delta \to \widehat{BG}[1/q]$ is a homotopy equivalence.

This justifies our thinking of BG^δ as a model for BG. Of course, the use of completions rather than localizations is essential here since $\pi_{2i}BGL\bar{k}_q^+ = 0$ and $\pi_{2i+1}BGL\bar{k}_q^+ = Z[q^{-1}]/Z$ [59, p. 585]. We can now verify that kG^δ represents the completion away from q of real or complex connective K-theory and that the equivalence $\hat{\lambda}$ is an infinite loop map.

<u>Theorem 2.8.</u> There is a unique isomorphism $\Lambda: kG^\delta \to \widehat{kG}[1/q]$ of ring spectra in $H\mathscr{S}$ such that the 0-component of the zeroth map of Λ is equivalent to $\hat{\lambda}: BG^\delta \to \widehat{BG}[1/q]$, G = O or G = U.

<u>Proof.</u> $\pi_0(kG_0^\delta)$ and $\pi_0(\widehat{kG}[1/q]_0)$ are both canonically isomorphic to the ring $\hat{Z}[1/q] = \underset{p \neq q}{\times} \hat{Z}_{(p)}$, and there is a unique (continuous) isomorphism of rings from one to the other. Denote this isomorphism by $\pi_0\hat{\lambda}$. By II.3.10 and a trivial diagram chase from Lemma 2.4, $\pi_0\hat{\lambda}$ and $\hat{\lambda}: BG^\delta \to \widehat{BG}[1/q]$ together determine an equivalence $\hat{\lambda}: kG_0^\delta \to \widehat{kG}[1/q]_0$ of ring spaces. Write b^δ for the composite

$$S^d \xrightarrow{b} BG \xrightarrow{\gamma} \widehat{BG}[1/q] \xrightarrow{\hat{\lambda}^{-1}} BG^\delta .$$

Thus $\hat{\lambda}_*[b^\delta] = [b]$. Let $\beta^\delta: kG^\delta \to \Omega^d kG^\delta$ be adjoint to the composite

$$kG^\delta \wedge S^d \xrightarrow{1 \wedge b^\delta} kG^\delta \wedge BG^\delta \xrightarrow{\phi} kG^\delta .$$

Then $(kG^\delta, \beta^\delta)$, $(\widehat{kG}[1/q], \beta)$, and the map $\hat{\lambda}$ of zeroth spaces satisfy the hypotheses of II.3.14 (and II.3.9). The conclusion follows from those results and II.3.2.

The following addendum is the reason that this result is of topological interest; it shows that the Frobenius automorphisms ϕ^r and Adams operations ψ^r (both completed away from q) agree under Λ.

<u>Theorem 2.9.</u> The following diagram commutes in $H\mathscr{S}$, G = O or G = U:

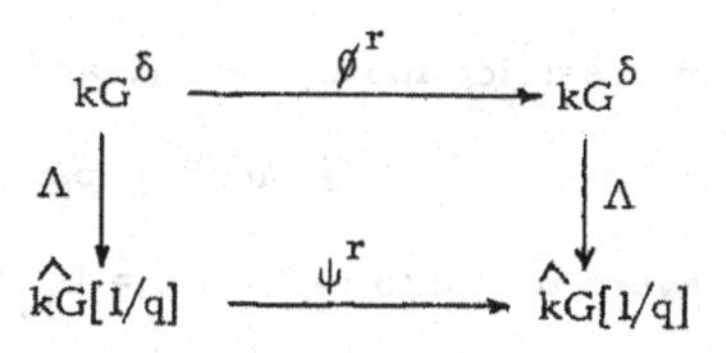

Proof. ϕ^r is induced by passage from bipermutative categories to E_∞ ring spaces to E_∞ ring spectra to completions away from q. Lemma 2.5 implies that the two composites induce the same map in $h\mathcal{V}$ on zero^{th} spaces and that this map is one of ring spaces. The conclusion follows by II.3.15 and the uniqueness clause of II.3.14.

Let $BG^\delta_\otimes$ denote the 1-component of the zero^{th} space of kG^δ. Clearly $O(1,\bar{k}_q) = Z_2$. $GL(1,\bar{k}_q) = \bar{k}^*_q$, and the completion away from q of the infinite loop map $B\mu : B\bar{k}^*_q \to B\mathbb{C}^*$ is an equivalence by a simple homological calculation. Let $BSG^\delta_\otimes$ denote the simply connected cover of $BG^\delta_\otimes$. The same proof as that of V.3.1 yields the following observation.

Lemma 2.10. $BO^\delta_\otimes$ and $BU^\delta_\otimes$ are equivalent as infinite loop spaces to $BO(1) \times BSO^\delta_\otimes$ and to $\hat{BU}(1)[1/q] \times BSU^\delta_\otimes$.

The 1-component $BG^\delta_\otimes \to BG_\otimes$ of the zero^{th} map of $\Lambda: kG^\delta \to \hat{kG}[1/q]$ is clearly compatible with the splittings given by the lemma and V.3.1. It is therefore an infinite loop map by the following theorem.

Theorem 2.11. $\Lambda: BSG^\delta_\otimes \to \widehat{BSG}[1/q]$ is an infinite loop map, $G = O$ or $G = U$.

Proof. By Theorem 1.6, it suffices to prove that the completion of Λ at each prime $p \neq q$ is an H^p_∞-map, and this will hold if it does so on the localized level with SG replaced by G. For clarity of notation, we treat the case $G = U$. The only additional point needed in the real case is that the relevant representation theoretical constructs, in particular the decomposition homomorphism, restrict appropriately, and the requisite information is contained in the appendix of Quillen's paper [58]. Let M be the the monoid of positive integers prime to p, let $X = B_M \mathcal{G}\mathcal{L}\bar{k}_q$, let $Y = \Gamma(X, \xi)$, let $\iota : X \to Y$ be the natural E_∞ map, and let $Z = \Gamma(B_M \mathcal{G}\mathcal{L}\mathbb{C}, \xi)$. Of course, $\mathcal{G}\mathcal{L}\mathbb{C}$ and $\mathcal{U}$ can be used interchangeably here, $\mathcal{G}\mathcal{L}\mathbb{C}$ being given its usual topology. By VII. 5.3 (and the diagram in its statement), we have infinite loop maps

$$\phi : \Gamma_M(B\mathcal{G}\mathcal{L}\bar{k}_q, \theta) \to Y \quad \text{and} \quad \phi : \Gamma_M(B\mathcal{G}\mathcal{L}\mathbb{C}, \theta) \to Z$$

which restrict on components to localizations at p. We agree to write $BU \times \{m\} = \Gamma_m(B\mathcal{G}\mathcal{L}\mathbb{C}, \theta)$ for $m \in M$; thus Z_m is the localization of $BU \times \{m\}$ at p. There is a unique map $f : Y_1 \to Z_1$ such that $f \circ \phi \simeq \phi \circ (\tilde{\lambda} * 1)$, where $\tilde{\lambda} * 1$ is the translation to 1-components of the map $\tilde{\lambda}$ of (**); f is an H-map since $\tilde{\lambda} * 1$ is so by the second diagram of (A). Since the map $\Lambda : BU_{\otimes}^{\delta} \to \widehat{BU}_{\otimes}[1/q]$ is the completion of $\tilde{\lambda} * 1$ away from q, it clearly suffices to prove that $f : Y_1 \to Z_1$ is an H_∞^p-map. By Remarks 1.4, f extends to an H-map $Y \to Z$ and it suffices to prove that the composite $f \circ \iota : X \to Z$ is an H_∞^p-map. The Brauer lifts of the inclusions $GL(m, k_r) \to GL(m, \bar{k}_q)$ fit together to give a unique map $\beta : BGL(m, \bar{k}_q) \to BU \times \{m\}$, and we also write β for its composite with the classifying map of any representation $H \to GL(m, \bar{k}_q)$. It follows easily from the definition of $\tilde{\lambda}$ and the proof of VII. 5.3 that $f \circ \iota \simeq \phi \circ \beta$: $X_m \to Z_m$. Thus we must prove that the following diagram is homotopy commutativ[e]

$$\begin{array}{ccc}
W \times_\pi BGL(m, \bar{k}_q)^p \simeq B(\pi \int GL(m, \bar{k}_q)) & \xrightarrow{B\tilde{c}_p} & BGL(m^p, \bar{k}_q) \\
\downarrow 1 \times \beta^p & & \downarrow \beta \\
W \times_\pi (BU \times \{m\})^p & & BU \times \{m^p\} \\
\downarrow 1 \times \phi^p & & \downarrow \phi \\
W \times_\pi (Z_m)^p & \xrightarrow{\xi} & Z_{m^p}
\end{array}$$

By an obvious limit argument, it suffices to prove this with $\bar{k}_q$ replaced by k_r (for each $r = q^a$) in the top row. By Lemma 1.1 and a transfer argument, the resulting diagram will homotopy commute if it does so with domain restricted to $W \times_\pi (BH)^p$ for a p-Sylow subgroup H of $GL(m, k_r)$ (since the index of $\pi \int H$ in $\pi \int GL(m, k_r)$ is prime to p and since Z_{m^p} is a p-local space). Let ℓ be a finite field between k_r and $\bar{k}_q$. By a trivial diagram chase, it suffices to prove that the diagram above homotopy commutes after replacement of $\bar{k}_q$ by ℓ in the top row and restriction of the domain to $W \times_\pi (BH)^p$. Let p^e be the maximal order of an element of $\pi \int H$. Construct a field $K \subset \mathbb{C}$ which contains all $(p^e)^{\underline{th}}$ roots of unity and has a discrete valuation whose valuation ring A has quotient field ℓ, $k_r \subset \ell \subset \bar{k}_q$. Consider the following diagram, where ρ is the inclusion of H in $GL(m, \ell)$, $r: \mathcal{GL}A \to \mathcal{GL}\ell$ is induced by the quotient map $A \to \ell$, and $i: \mathcal{GL}A \to \mathcal{GL}\mathbb{C}$ is induced by the inclusion $A \to \mathbb{C}$:

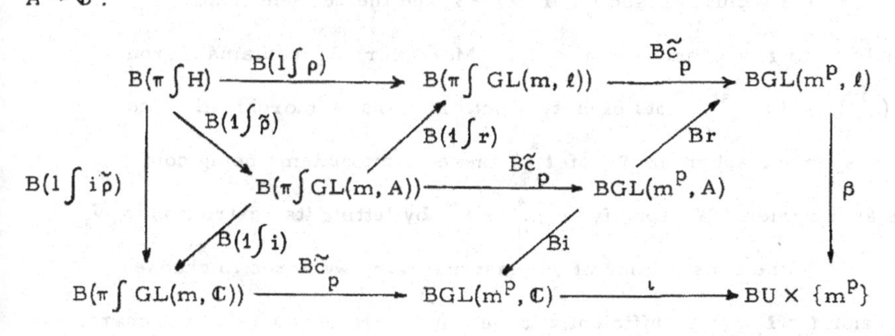

The two interior parallelograms obviously commute. Let $\hat{K}$ be the completion of K. Then $R_K H = R_{\hat{K}} H$ [69, 14.6], and the decomposition homomorphism $d: R_{\hat{K}} H \to R_\ell H$ is an isomorphism [69, 15.2 and 15.5]. Thus $d: R_K H \to R_\ell H$ is also an isomorphism, and, since d maps $R_K^+ H$ onto $R_\ell^+ H$, any representation $\rho: H \to GL(m, \ell)$ lifts to a representation $\tilde{\rho}: H \to GL(m, A)$ [69, especially the remarks on pp. 136, 139, 141]. This lifting to an honenst rather than virtual representation is crucial. By Proposition 1.2 (applied to $\mathcal{B}\mathcal{L}C$ under $\otimes$), the fact that ϕ is an infinite loop map, and a trivial diagram chase, it suffices to prove that the composite $BH \xrightarrow{B(i\tilde{\rho})} BGL(m, \mathbb{C}) \xrightarrow{\iota} BU \times \{m\}$ is homotopic to β and that the outer rectangle of the diagram is homotopy commutative. Actually, because of the homomorphism $\mu: \bar{k}_q^* \to \mathbb{C}^*$ in our definition of Brauer lifting, these assertions will in general be off by certain Adams operations. To rectify this, we need only choose μ consistently with the requisite decomposition homomorphism (in a manner independent of p). We do this as follows. Let $\mathfrak{Q}_0 = (q)$ in $B_0 = Z$ and let $A_0 = Z_{(q)}$. Inductively, given $\mathfrak{Q}_{j-1}, B_{j-1}$, and A_{j-1}, let A_j be the localization of the ring of cyclotomic integers $B_j = Z[\exp(2\pi i/(q^j-1)(q^{j-1}-1)\cdots(q-1)]$ at a chosen prime ideal $\mathfrak{Q}_j$ which contains $\mathfrak{Q}_{j-1} \subset B_{j-1}$. Let $K_j \subset \mathbb{C}$ be the field of fractions of A_j, let ℓ_j be the quotient of A_j by its maximal ideal, let $i: A_j \to \mathbb{C}$ be the inclusion, and let $r: A_j \to \ell_j$ be the quotient map. Obviously $\operatorname{char} \ell_j = q$ and $\varinjlim \ell_j = \bar{k}_q$. Moreover, A_j contains a group ν_j of $(q^j-1)\cdots(q-1)^{st}$ roots of unity which r maps isomorphically onto the corresponding subgroup $\bar{\nu}_j$ of ℓ_j^*, these isomorphisms being compatible as j varies. We specify $\mu: \bar{k}_q^* \to \mathbb{C}^*$ by letting its restriction to $\bar{\nu}_j$ be $i \circ r^{-1}$. In the construction of our last diagram, we agree to choose $A = A_j$ and $\ell = \ell_j$ for j sufficiently large. It is then obvious that the character of $i\tilde{\rho}: H \to GL(m, \mathbb{C})$ is χ_p, so that $\iota \circ B(i\tilde{\rho}) \simeq \beta$. Similarly, if G is a finite p-group with no elements of order greater than p^e and if $\sigma: G \to GL(m^p, A)$ is a representation, then the character of $i\sigma: G \to GL(m^p, \mathbb{C})$ is $\chi_{r\sigma}$ and $\iota \circ B(i\sigma): BG \to BU \times \{m^p\}$ is therefore homotopic to β. With $\sigma = \tilde{c}_p \circ (1 \int \tilde{\rho}): H \to GL(m^p, A)$, it follows that the outer rectangle of our diagram is indeed homotopy commutative.

§3. Finite fields, Frobenius, and B Coker J

Throughout this section, all spaces and spectra are to be completed at a fixed prime p and $r = q^a$ (q odd) is to be $r(p)$. Thus $r = 3$ if $p = 2$ and r reduces mod p^2 to a generator of the group of units of Z_{p^2} if $p > 2$. We retain the notations of the previous section and continue with the discussion of discrete models for various of the spaces and maps in the J-theory diagram of V §3.

We have an equivalence of orientation sequences

$$\begin{array}{ccccccc}
SF & \xrightarrow{e} & BO^{\delta}_{\otimes} & \xrightarrow{\tau} & B(SF;kO^{\delta}) & \xrightarrow{q} & BSF \\
\| & & \downarrow{\scriptstyle \Lambda} & & \downarrow{\scriptstyle B\Lambda} & & \| \\
SF & \xrightarrow{e} & BO_{\otimes} & \xrightarrow{\tau} & B(SF;kO) & \xrightarrow{q} & BSF
\end{array}$$

(compare V.2.4) and an equivalence of fibration sequences

$$\begin{array}{ccccccc}
Spin^{\delta}_{\otimes} & \longrightarrow & BC^{\delta}_{p} & \longrightarrow & B(SF;kO^{\delta}) & \xrightarrow{c(\phi^r)} & BSpin^{\delta}_{\otimes} \\
\downarrow{\scriptstyle \Omega\Lambda} & & \downarrow & & \downarrow{\scriptstyle B\Lambda} & & \downarrow{\scriptstyle \Lambda} \\
Spin_{\otimes} & \longrightarrow & BC_{p} & \longrightarrow & B(SF;kO) & \xrightarrow{c(\psi^r)} & BSpin_{\otimes}
\end{array}.$$

Here $BSpin^{\delta}_{\otimes}$ is the 2-connected cover of $BSO^{\delta}_{\otimes}$ and $Spin^{\delta}_{\otimes}$ is its loop space. $c(\phi^r)$ is the universal cannibalistic class (defined above V.2.2) determined by $\phi^r : kO^{\delta} \to kO^{\delta}$ and is an infinite loop map because ϕ^r is the completion of a map of E_{∞} ring spectra. The fibre BC^{δ}_{p} of $c(\phi^r)$ is thus an infinite loop space, and we think of it as BC_p endowed with an infinite loop space structure. We shall prove in Theorem 3.4 that both diagrams above are commutative diagrams of infinite loop spaces and maps.

In order to obtain a better understanding of the infinite loop space BC^{δ}_{p}, we construct discrete models for the spectra j_p and jO_2 of V.5.16. Recall the functor $\Omega^{\infty}T$ from E_{∞} ring spaces to E_{∞} ring spectra of VII.4.1.

Definition 3.1. Define $j_2^\delta = \Omega^\infty TB\mathcal{N}k_3$ and $jO_2^\delta = \Omega^\infty TB\mathcal{O}k_3$. For $p > 2$, define $j_p^\delta = \Omega^\infty TB\mathcal{GL}k_r$. The bipermutative categories $\mathcal{N}k_3$, $\mathcal{O}k_3$, and $\mathcal{GL}k_r$ are specified in VI. 5.7, 5.3, and 5.2 (and the specified E_∞ ring spectra are understood to be completed at p). Let J_p^δ and $J_{\otimes p}^\delta$ denote the 0-component and 1-component of the zero[th] space of j_p^δ (which is equivalent to $J_p^\delta \times \hat{Z}_{(p)}$), and let JO_2^δ and $JO_{\otimes 2}^\delta$ denote the 0-component and 1-component of $(jO_2^\delta)_0$.

The following theorem is based on ideas and results of Quillen [57, 59] and Fiedorowicz and Priddy [28].

Theorem 3.2. There are equivalences $\nu: j_p^\delta \to j_p$ and $\bar{\nu}: jO_2^\delta \to jO_2$ such that the following diagrams commute in $H\mathscr{S}$:

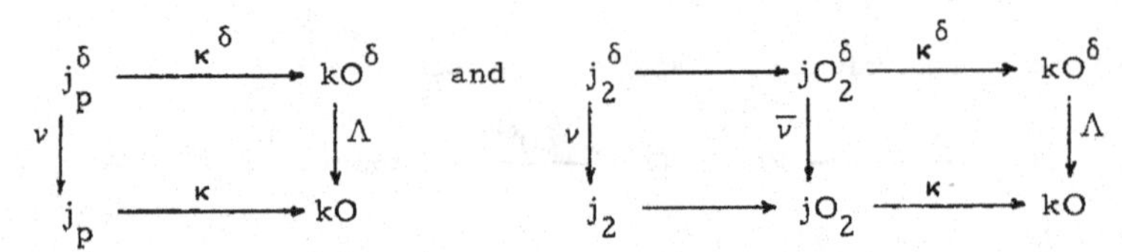

where the κ^δ are induced by inclusions of bipermutative categories (when $p > 2$, in the sense that $j_p^\delta \xrightarrow{\kappa^\delta} kO^\delta \longrightarrow kU^\delta$ is induced by $\mathcal{GL}k_r \to \mathcal{GL}\bar{k}_q$).

Proof. It will be convenient to treat the cases $p > 2$ and $p = 2$ separately. We adopt the obvious discrete models analog of the notations in V. 5.14.

(i) $p > 2$. In view of Theorem 2.9 (and [48, I (2.12)]), we have the following comparisons of fibration sequences in $H\mathscr{S}$, where $F\psi^r$ and $F\phi^r$ denote the relevant fibres:

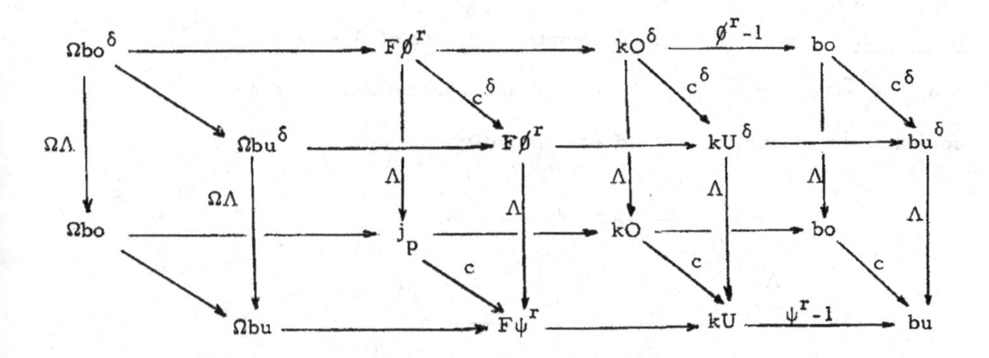

Since $p > 2$, $bo \simeq bspin$. The arrows labelled c^δ are induced by passage to completed spectra from the inclusion $\mathcal{O}\bar{k}_q \to \mathcal{GL}\bar{k}_q$ of bipermutative categories. By [1, 5.1] and Theorem 2.8, the dotted arrows are all equivalences in $H\mathcal{S}$. On the level of bipermutative categories, $\phi^r: \mathcal{GL}\bar{k}_q \to \mathcal{GL}\bar{k}_q$ restricts to the identity on $\mathcal{GL}k_r$. By passage to completed spectra, we conclude that the composite of $\phi^r - 1$ with the map $j_p^\delta \to kU^\delta$ induced by the inclusion of $\mathcal{GL}k_r$ in $\mathcal{GL}\bar{k}_q$ is trivial. There results a lift $\mu: j_p^\delta \to F\phi^r$, and μ obviously induces an isomorphism on π_0. Since $KU^{-1}BG = 0$ for a finite group G [14, 4.2] and since the zeroth space functor commutes with fibres [48, VIII], the 0-component of the zeroth map of μ is determined by the homotopy commutative diagram

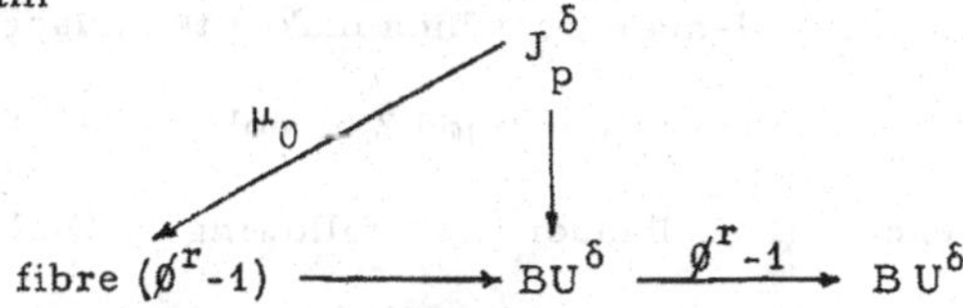

Quillen [59, p. 576] proved that μ_0 induces an isomorphism on mod p homology and is therefore a homotopy equivalence (since J_p^δ and the fibre of $\phi^r - 1$ are p-complete simple spaces). It follows that μ induces isomorphisms on π_i for all i and is thus an equivalence. The desired equivalence $\nu: j_p^\delta \to j_p$ is $c^{-1} \circ \Lambda \circ \mu$.

(ii) $p = 2$. Let $FO\phi^3$ and $F\phi^3$ denote the fibres of $\phi^3 - 1 : kO^\delta \to bso^\delta$ and of $\phi^3 - 1 : kO^\delta \to bspin^\delta$. By V.5.15 and Theorems 2.8 and 2.9, comparisons of fibrations yield a commutative diagram

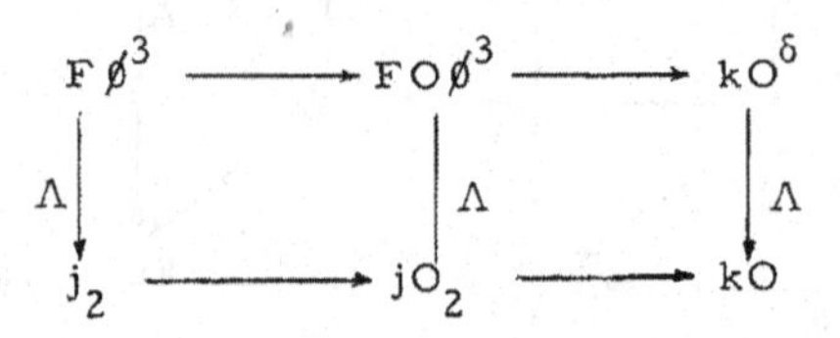

in which the maps Λ are equivalences. The composite of $\phi^3 - 1 : kO^\delta \to bso^\delta$ with $\kappa^\delta : jO_2^\delta \to kO^\delta$ is trivial since $\phi^3 : \mathcal{O}\bar{k}_3 \to \mathcal{O}\bar{k}_3$ restricts to the identity on $\mathcal{O}k_3$ and since $[jO_2^\delta, bso^\delta] \cong [jO_2^\delta, bo^\delta]$ by the proof of V.5.15 and the fact that $H^0 jO_2^\delta = H^0 kO^\delta$ (where H denotes mod 2 cohomology). There results a lift $\bar{\mu} : jO_2^\delta \to FO\phi^3$, and $\bar{\mu}$ obviously induces an isomorphism on π_0. Restriction to the 0-component of zeroth spaces gives a homotopy commutative diagram

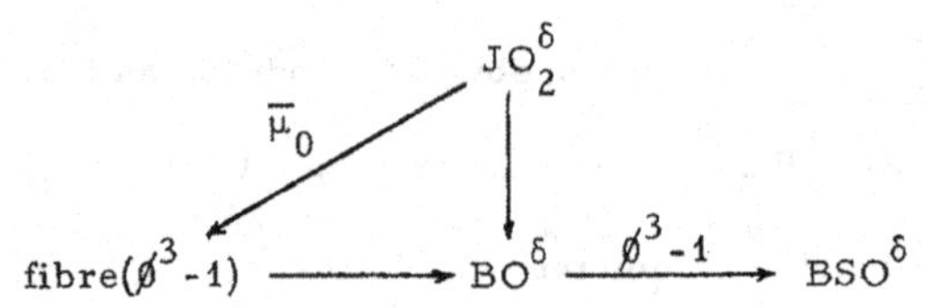

Here $\bar{\mu}_0$ is not determined by the diagram, but Fiedorowicz and Priddy [28] have proven that any H-map $\bar{\mu}_0$ which makes the triangle homotopy commute induces an isomorphism on mod 2 homology and is therefore a homotopy equivalence. (Friedlander [29], following up Quillen's ideas about étale cohomology [57], earlier obtained a particular equivalence $\bar{\mu}_0$, not necessarily an infinite loop map.) Thus $\bar{\mu}$ is an equivalence in $H\mathcal{L}$. Next, consider the following diagram in $H\mathcal{L}$:

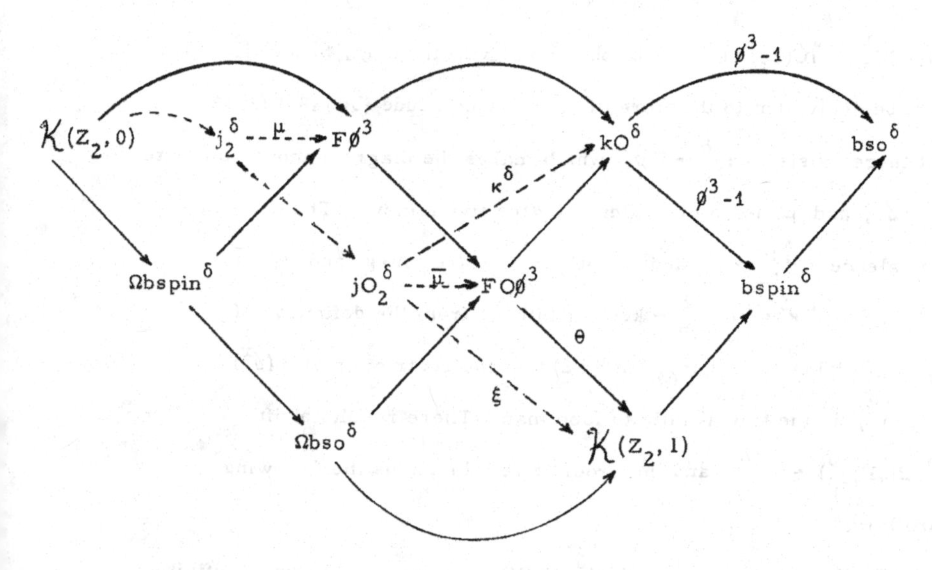

The right triangle commutes and induces θ in such a manner that $F\phi^3$ is canonically equivalent to the fibre of θ and the solid arrow diagram is a braid of fibrations (by [48, I (2.13)]. θ restricts non-trivially to Ωbso^δ, and we define $\xi = \theta\bar{\mu}$. We need a slight calculation to construct μ. Recall that $H^*bso \cong \Sigma^2(A/ASq^3)$ (e.g., by [8]). Since $H^2kO^\delta = 0$, $\phi^3-1: kO^\delta \to bso^\delta$ induces the trivial map on H^* and we have an exact sequence

$$0 \to A/ASq^1 + ASq^2 \to H^*FO\psi^3 \to \Sigma(A/ASq^3) \to 0 .$$

Thus $H^1FO\phi^3 = Z_2$ and its unique non-zero class θ restricts to the generator of $H^*\Omega bso^\delta$. By inspection of the fibration $F\pi \to jO_2^\delta \xrightarrow{\pi} \mathcal{K}(\hat{Z}_{(2)},0)$ with 0-connected fibre, we see that this is consistent with the known fact (e.g. [28]) that $H^1JO_2^\delta = H^1BOk_3$ is $Z_2 \oplus Z_2$ with non-zero classes corresponding to the determinant, the spinor norm, and their product; we denote the last of these classes by ξ_0. In view of VI. 5.3 and VI. 5.7, J_2^δ is equivalent to the fibre of $\xi_0: JO_2^\delta \to K(Z_2,1)$ (see [28]). Thus the cofibre of $j_2^\delta \to jO_2^\delta$ is $\mathcal{K}(Z_2,1)$ by the long exact homotopy sequence. Clearly, the cofibre map $jO_2^\delta \to K(Z_2, 1)$ must be the non-trivial map ξ, hence j_2^δ must be equivalent to the fibre of ξ. We conclude (by [48, I (2.12)] that there exists $\mu: j_2^\delta \to F\phi^3$ which makes the diagram above commute in $H\mathscr{J}$, and μ is an equivalence by the five lemma. The desired equivalences $\nu: j_2^\delta \to j_2$ and $\bar{\nu}: jO_2^\delta \to jO_2$ are $\Lambda \cdot \mu$ and $\Lambda \cdot \bar{\mu}$.

Since $\phi^r \kappa^\delta = \kappa^\delta: j_p^\delta \to kO^\delta$, it follows from the definition of $c(\phi^r): B(SF; kO^\delta) \to BSpin_\otimes^\delta$ (in V § 2) that the restriction of $c(\phi^r)$ to $B(SF; j_p^\delta)$ is the trivial infinite loop map. There results a lift $\zeta^\delta: B(SF; j_p^\delta) \to BC_p^\delta$, and the proof of V.5.17 yields the following corollary.

<u>Corollary 3.3</u>. $\zeta^\delta: B(SF; j_p^\delta) \to BC_p^\delta$ is an equivalence of infinite loop spaces.

In V. § 5, j_p was regarded as a ring spectrum by pullback along ν^{-1}. On 1-components of zeroth spaces, ν and $\bar{\nu}$ restrict to composite equivalences

$$J_\otimes^\delta \xrightarrow{\mu_0} F\phi_\otimes^r \xrightarrow{\Lambda} J_{\otimes p} \quad \text{and} \quad JO_{\otimes 2}^\delta \xrightarrow{\bar{\mu}_0} FO\phi_\otimes^\delta \xrightarrow{\Lambda} JO_{\otimes 2}$$

where $F\phi_\otimes^\delta$ and $FO\phi_\otimes^\delta$ denote the fibres of $\phi^r/1: BO_\otimes^\delta \to BSpin_\otimes^\delta$ and of $\phi^3/1: BO_\otimes^\delta \to BSO_\otimes^\delta$. We shall see in the following theorem that the maps Λ may be regarded as infinite loop maps in view of Theorem 2.11. When $p > 2$, μ_0 is easily seen to be an H-map; when $p = 2$, not even this much is clear in view of the non-uniqueness of $\bar{\mu}_0$. However, the proof of the following theorem will yield a possibly different (when $p = 2$) map $\mu_\otimes: J_{\otimes p}^\delta \to F\phi_\otimes^r$ which is an equivalence of infinite loop spaces, and an analogous argument gives an equivalence $\bar{\mu}_\otimes: JO_{\otimes 2}^\delta \to FO\phi_\otimes^3$ of infinite loop spaces. The composite $\Lambda\mu_\otimes$ plays a central role in the "multiplicative Brauer lift diagram" displayed on the following page.

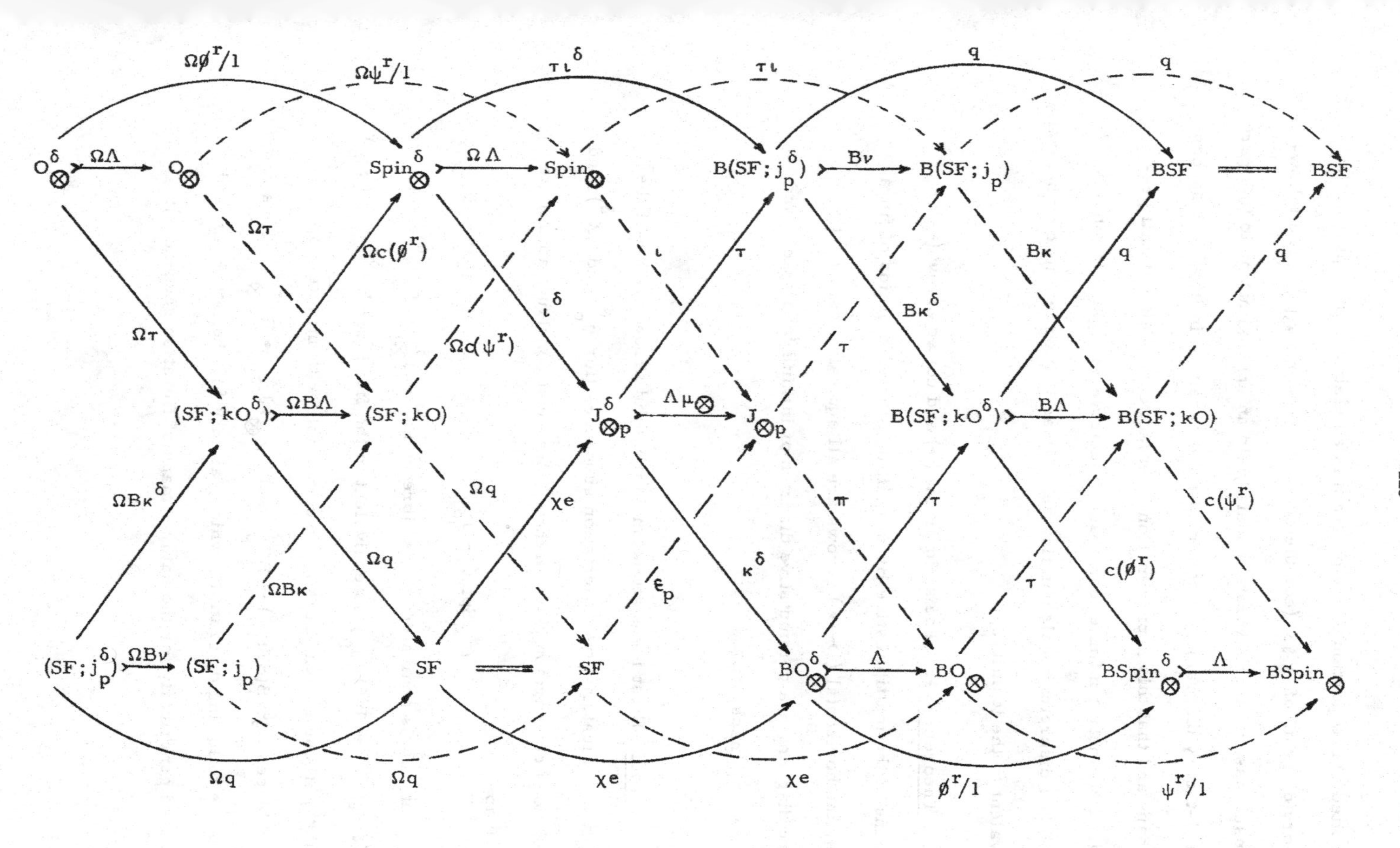

$\Omega\phi^r/1$
$\Omega\psi^r/1$
$\tau\iota^\delta$
$\tau\iota$
q
q
$O^\delta_\otimes$
$\Omega\Lambda$
$O_\otimes$
$Spin^\delta_\otimes$
$\Omega\Lambda$
$Spin_\otimes$
$B(SF; j_p^\delta)$
$B\nu$
$B(SF; j_p)$
BSF
BSF
$\Omega\tau$
$\Omega c(\phi^r)$
ι
τ
$B\kappa$
q
q
$\Omega\tau$
ι^δ
$\Omega c(\psi^r)$
$B\kappa^\delta$
τ
$(SF; kO^\delta)$
$\Omega B\Lambda$
$(SF; kO)$
$J^\delta_{\otimes p}$
$\Lambda\mu_\otimes$
$J_{\otimes p}$
$B(SF; kO^\delta)$
$B\Lambda$
$B(SF; kO)$
$\Omega B\kappa^\delta$
Ωq
χe
π
τ
$c(\psi^r)$
Ωq
κ^δ
$c(\phi^r)$
$\Omega B\kappa$
ϵ_p
τ
$(SF; j_p^\delta)$
$\Omega B\nu$
$(SF; j_p)$
SF
SF
$BO^\delta_\otimes$
Λ
$BO_\otimes$
$BSpin^\delta_\otimes$
Λ
$BSpin_\otimes$
Ωq
Ωq
χe
χe
$\phi^r/1$
$\psi^r/1$

The dotted arrow portion of the diagram is an elaboration of part of the J-theory diagram of V §3 (completed at p, with $r = r(p)$). The following result asserts that discrete models yield an approximation to this part of the J-theory diagram by a commutative diagram of infinite loop spaces and maps and that this approximation is in fact consistent with all preassigned geometric infinite loop space structures in sight. In other words, our ad hoc discrete models notation behaves as if it were a functor naturally equivalent to the identity.

Theorem 3.4. The solid arrow ($\rightarrow$) and dotted arrow ($\dashrightarrow$) portions of the multiplicative Brauer lift diagram are braids of fibrations, the horizontal ($\rightarrowtail$) arrows are all equivalences, and the entire diagram is a commutative diagram of infinite loop spaces and infinite loop maps.

Proof. First focus attention on the solid arrow portion of the diagram. It features two orientation sequences (for j_p^δ and kO^δ) and the obvious comparison between them. We must construct an infinite loop map

$$\iota^\delta : Spin_\otimes^\delta \rightarrow J_{\otimes p}^\delta$$

such that ι^δ is equivalent to the fibre $\pi: F\kappa^\delta \rightarrow J_{\otimes p}^\delta$ of κ^δ and $\tau\iota^\delta: Spin_\otimes^\delta \rightarrow B(SF; j_p^\delta)$ is equivalent to the fibre $\pi: FB\kappa^\delta \rightarrow B(SF; j_p^\delta)$ of $B\kappa^\delta$, these equivalences being compatible with infinite loop equivalences $\zeta^\delta: B(SF; j_p^\delta) \rightarrow BC_p^\delta$ and $\mu_\otimes: J_{\otimes p}^\delta \rightarrow F\phi_\otimes^r$. Thus consider the following diagram (in which, as in V §3, the letters π and ι are used generically for the natural maps of fibration sequences):

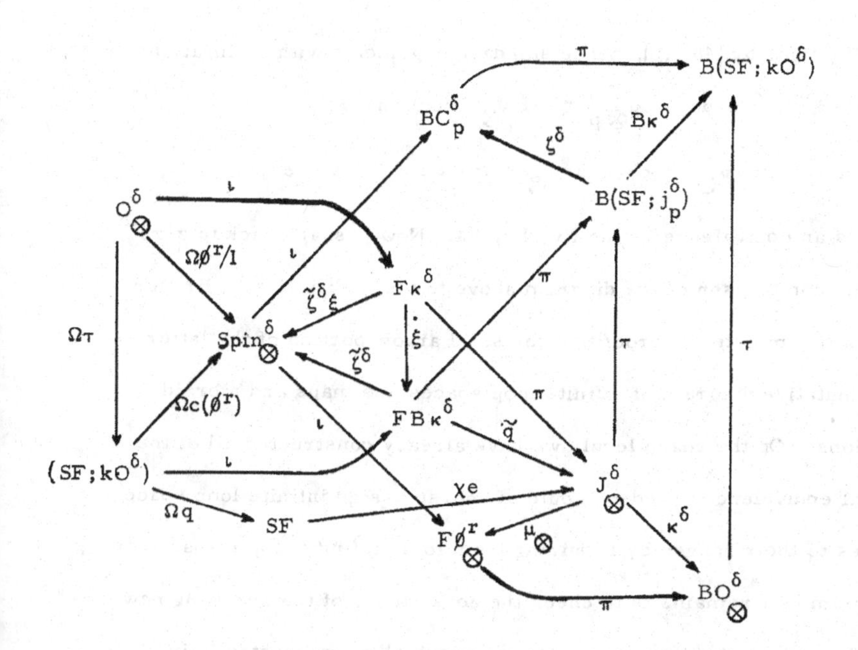

$\tau: J^\delta_{\otimes p} \to B(SF; j^\delta_p)$ is equivalent to the fibre of $q: B(SF; j^\delta_p) \to BSF$, and $q: B(SF; kO^\delta) \to BSF$ induces $\tilde{q}: FB\kappa^\delta \to J^\delta_{\otimes p}$ (by base change). By [48, I.(2.13)], (which is a precise form of Verdier's axiom for fibrations), there is a canonical equivalence $\xi: F\kappa^\delta \to FB\kappa^\delta$ such that

$$\xi \circ \iota \simeq \iota \circ \Omega\tau \quad \text{and} \quad \tilde{q} \circ \xi \simeq \pi .$$

Clearly, ζ^δ induces an equivelence $\tilde{\zeta}^\delta : FB\kappa^\delta \to Spin^\delta_\otimes$ such that

$$\zeta^\delta \circ \pi \simeq \iota \circ \tilde{\zeta}^\delta \quad \text{and} \quad \tilde{\zeta}^\delta \circ \delta \simeq \Omega c(\emptyset^r) .$$

Define $\iota^\delta = \tilde{q}(\tilde{\zeta}^\delta)^{-1} : Spin^\delta_\otimes \to J^\delta_{\otimes p}$. It remains to construct $\mu_\otimes : J^\delta_{\otimes p} \to F\emptyset^r_\otimes$, and we note that

$$(\tilde{\zeta}^\delta \circ \xi) \circ \iota \simeq \tilde{\zeta}^\delta \circ \iota \circ \Omega\tau \simeq \Omega c(\emptyset^r) \circ \Omega\tau \simeq \Omega\emptyset^r/1 : O^\delta_\otimes \to Spin^\delta_\otimes .$$

The constructions so far all result by passage to zeroth spaces from the analogous constructions on spectra, hence we may regard the diagram as one of connective spectra in $H\mathcal{J}$. By abuse, we retain the notations of the diagram on the spectrum level. Here cofibrations and fibrations agree up to sign, by [48, XI], hence standard arguments with cofibrations show that $\tilde{\zeta}^{\delta}\circ\xi$ induces $\mu_{\otimes}: J^{\delta}_{\otimes p} \to F\phi^r_{\otimes}$ such that

$$\pi\circ\mu_{\otimes} = \kappa^{\delta} \text{ and } \mu_{\otimes}\circ\pi = \iota\circ\tilde{\zeta}^{\delta}\xi \quad \text{in } H\mathscr{J},$$

and $\mu_{\otimes}$ is an equivalence by the five lemma. Now passage back to zeroth spaces and comparison of the diagram above to the multiplicative Brauer lift diagram complete the proof that the solid arrow portion of the latter is a commutative diagram of infinite loop spaces and maps and a braid of fibrations. On the space level, we have already constructed all of the horizontal equivalences, and we could of course assign infinite loop space structures to their ranges by requiring them to be infinite loop maps. The problem that remains is to check the consistency of the resulting new infinite loop space structures with the geometrically constructed infinite loop space structures already existing on $BO_{\otimes}$, $BSpin_{\otimes}$, $B(SF;kO)$, and their loop spaces. By V.3.1, Lemma 2.10, and Theorem 2.11, we have that

$$\Lambda: BO^{\delta}_{\otimes} \to BO_{\otimes} \text{ and } \Lambda: BSpin^{\delta}_{\otimes} \to BSpin_{\otimes}$$

are both infinite loop maps. We may therefore specify $\psi^r/1$ as an infinite loop map by $\psi^r/1 = \Lambda\circ\phi^r/1\circ\Lambda^{-1}$ (compare V.7.6). The equivalence $\Lambda: F\phi^r_{\otimes} \to J_{\otimes p}$ obtained by comparison of fibrations is then an infinite loop map if $J_{\otimes p}$ is given an infinite loop space structure as the fibre of $\psi^r/1$. Next, specify $\varepsilon_p = \Lambda\mu_{\otimes}\circ\chi e: SF \to J_{\otimes p}$ as an infinite loop map. On the space level, parts of the multiplicative Brauer lift diagram already known to commute then imply that

$$\varepsilon_p\circ\Omega q \simeq \iota\circ\Omega c(\psi^r): (SF;kO) \to J_{\otimes p} \text{ and } \pi\circ\varepsilon_p \simeq \chi e: SF \to BO_{\otimes}.$$

These were the defining conditions for the map labelled ε_p in V§3. We have that $\pi\varepsilon_p = \chi e = \Lambda\chi e: SF \to BO_{\otimes}$ as infinite loop maps in view of V.7.9. Delooping once, we conclude (by [48, I(2.12)]) that there is an infinite loop map $B'\Lambda: B(SF;kO^{\delta}) \to B(SF;kO)$ such that

$$q \circ B'\Lambda = q: B(SF;kO^{\delta}) \to BSF \text{ and } \tau \circ \Lambda = B'\Lambda \circ \tau: BO^{\delta}_{\otimes} \to B(SF;kO)$$

as infinite loop maps. We must verify that $B'\Lambda$ is homotopic to $B\Lambda$, and of course $q \circ B\Lambda \simeq q$ and $\tau \circ \Lambda \simeq B\Lambda \circ \tau$ on the space level. Thus $B\Lambda/B'\Lambda$ factors as $\tau\omega$ for some map $\omega: B(SF;kO^{\delta}) \to BO_{\otimes}$, and $\tau\omega\tau \simeq *$. Since $B(SF;kO^{\delta})$ has the homotopy type of $BSpin \times BC_p$, by V.4.7 and V.4.8, and since $[BC_p, BO_{\otimes}] = 0$, by V.7.8, ω may be regarded as a map $BSpin \to BO_{\otimes}$. It clearly induces the trivial homomorphism on rational (indeed, on integral) cohomology, and V.2.8 and 2.10 imply that it is null homotopic. Thus $B\Lambda \simeq B'\Lambda$. We may now specify $c(\psi^r)$ as an infinite loop map by $c(\psi^r) = \Lambda \circ c(\phi^r) \circ (B\Lambda)^{-1}$. Similarly, we specify $B(SF;j_p)$ as an infinite loop space by requiring $B\nu$ to be an infinite loop map and we specify $B\kappa = B\Lambda \circ B\kappa^{\delta} \circ (B\nu)^{-1}: B(SF;j_p) \to B(SF;kO)$ and $\tau = B\nu \circ \tau \circ (\Lambda\mu_{\otimes})^{-1}: J_{\otimes p} \to B(SF;j_p)$ as infinite loop maps. The remaining verifications are trivial.

We single out the following part of the theorem for emphasis (compare V.5.13).

<u>Corollary 3.5.</u> The composite $SF \xrightarrow{\chi e} J^{\delta}_{\otimes p} \xrightarrow{\mu_{\otimes}} F\phi^r_{\otimes} \xrightarrow{\Lambda} J_{\otimes p}$ may be taken as the map $\varepsilon_p: SF \to J_{\otimes p}$ of the J-theory diagram.

The force of this assertion lies mainly at the prime 2. At odd primes it is almost trivial, since there we have

$$[SF, SO_{\otimes}] \cong [Q_0S^0, SO_{\otimes}] \cong [B\Sigma_{\infty}, SO_{\otimes}] = 0,$$

by VII.3.4 and [14], so that ε_p is uniquely determined by the fact that its composite with $\pi: J_{\otimes p} \to BO_{\otimes}$ is homotopic to $\chi e: SF \to BO_{\otimes}$.

We digress to give the following application of the corollary, which summarizes Quillen's results [60] about K_*Z.

Remarks 3.6. For any commutative topological ring A, we have a commutative diagram of bipermutative categories

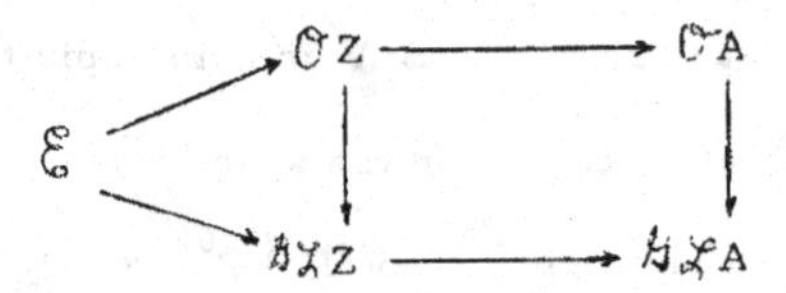

and a derived commutative diagram of K-groups in positive degrees

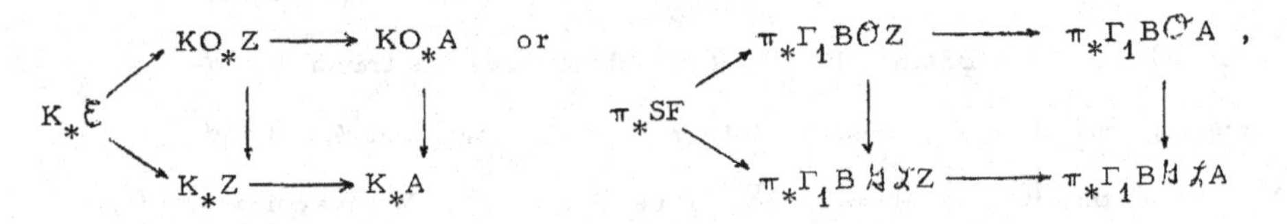

where the second diagram results from the first by translation from 0-components to 1-components of zeroth spaces of spectra and where $\Gamma_1 B\mathcal{E}$ is identified with SF via VII.3.4, 4.4, and 4.5. By VII. 4.6, $K_*\mathcal{E} = \pi_*^s$ maps monomorphically onto a direct summand of KO_*Z, the complementary summand being isomorphic to $\pi_*^s(RP^\infty)$. When $A = \mathbb{R}$, $KO_*A = K_*A = \pi_*BO_\otimes$ and V.5.6 shows that the element $\mu_i \in \pi_i SF$, $i \equiv 1$ or 2 mod 8, defines a direct summand Z_2 in K_iZ. Let ${}_pJ_i$ denote the p-torsion in the image of $j_*: \pi_i Spin \to \pi_i SF$. When $p > 2$ and $A = k_{r(p)}$, the p-torsion subgroup of K_*A is isomorphic to ${}_pJ_*$ and, by Corollary 3.5 and V.4.6, ${}_pJ_*$ is a direct summand of K_*Z. Finally, consider ${}_2J_i$. The image of ${}_2J_i = Z_2$ in K_iZ is unknown, $i \equiv 0$ or 1 mod 8 (and $i \geq 8$). Let $A = k_3$. The 2-torsion subgroup of $KO_{4i-1}k_3 = K_{4i-1}k_3$ is ${}_2J_{4i-1}$. Write JU_2 and JU_2^δ for the fibre of $\psi^3 - 1: BU \to BU$ and for $\Gamma_0 B\mathcal{GL}k_3$ (completed at 2). By the proof of Theorem 3.2, there is an equivalence $\bar{\nu}_0: JU_2^\delta \to JU_2$ under which the natural map $JO_2^\delta \to JU_2^\delta$ corresponds to the map $c: JO_2 \to JU_2$ induced by complexification. By [1, 5.2], $\pi_{4i-1}JO_2$ and $\pi_{4i-1}JU_2$ are the same group and c_{4i-1} is the identity if i is even and multiplication by 2 if i is odd. Therefore, by Corollary 3.5 and V.4.6, ${}_2J_{8i-1}$ is a direct summand of K_*Z and the image in K_*Z of the element of order 2 in ${}_2J_{8i-5}$ maps to zero in K_*k_3. Quillen [60] proved that ${}_2J_{4i-1}$ maps monomorphically to $K_{4i-1}Z$ by noting that Adams' e-invariant can be identified with the map $\pi_{4i-1}SF \to \pi_{4i-1}X$ induced by the unique lift $\zeta: SF \to X$ of $\chi e: SF \to BO_{\otimes}$ to the fibre X of the Pontryagin character $BO_{\otimes} \to \underset{i\geq 1}{\times} K(Q, 4i)$ and observing that ζ necessarily factors through $\Gamma_1 B \mathcal{GL} Z$ because the Chern classes of representations of discrete groups are torsion classes. Karoubi [34] found that ${}_2J_3$ is not a direct summand of K_3Z, and Lee and Sczcarba [37] proved the deep result that K_3Z is exactly Z_{48}.

§4. The splitting of SF at odd primes

Again, all spaces and spectra are to be completed at a fixed prime $p \neq q$ and $r = q^a$ is to be $r(p)$. Actually, almost all spaces in sight will have finite homotopy groups, hence localization will agree with completion.

Theorem 3.4 focuses attention on the orientation sequence

$$SF \xrightarrow{e} J^{\delta}_{\otimes p} \xrightarrow{\tau} B(SF; j^{\delta}_{p}) \xrightarrow{q} BSF .$$

The map τ is null homotopic by the splitting of SF in V.4.6-4.8 (and Corollary 3.5). When $p = 2$, [70, 9.11 or 26 II.12.2] show that there is no splitting $SF \simeq C_2 \times J_2$ as H-spaces, and presumably the first delooping of τ already fails to be null homotopic. When $p > 2$, we shall prove an exponential law for $\mathcal{GL} k_r$ and shall use it to split SF and $B(SF; kO^{\delta})$ as infinite loop spaces; it will follow that τ is trivial as an infinite loop map.

Let M denote the monoid $\{r^n \mid n \geq 0\}$. Subscripts M will denote unions of components indexed on M. Since $Q_\infty S^0$ is the free spectrum generated by S^0 (by II.1.6) and also the free $\mathcal{Q}\times\mathcal{L}$ - spectrum generated by S^0 (by IV.2.4 and 2.5), there is an exponential unit map of spectra $e_r : Q_\infty S^0 \to \Omega^\infty T(Q_M S^0, \xi)$ specified on S^0 by $0 \to 1$ and $1 \to x$ for any chosen point $x \in Q_r S^0$ and also a unit map of $\mathcal{Q}\times\mathcal{L}$ -spectra $e: Q_\infty S^0 \to \Omega^\infty T(B\,\mathcal{GL}k_r, \theta)$. By VI.5.2 and 5.6, we have a unit functor $e: \mathcal{E} \to \mathcal{GL}k_r$ and a forgetful functor $f: \mathcal{GL}k_r \to \mathcal{E}$. Let $e_r = fe: \mathcal{E} \to \mathcal{E}$ (which is an exponential map of permutative categories) and let $g = ef: \mathcal{GL}k_r \to \mathcal{GL}k_r$. Recall from VI.5.1 that $B\mathcal{E} = DS^0$. With these notations, freeness and VII.4.4, 4.5, 5.3, and 5.4 yield the following homotopy commutative diagram in which all maps indicated by $\simeq$ are homotopy equivalences and $\alpha_p^\delta : J_p^\delta \to SF$ is defined to be the composite from the lower left to the upper right corner:

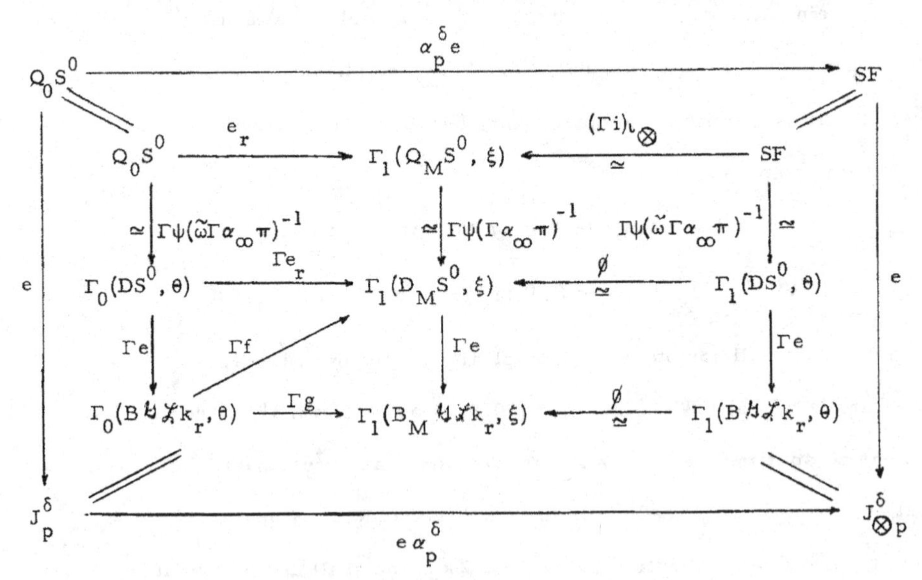

Here $(\Gamma i)\iota_\otimes$ and the maps $\emptyset$ are equivalences since they are localizations at M and we are further localizing or completing all spaces at p. The maps $e: Q_0S^0 \to J_p^\delta$ and $e: SF \to J_{\otimes p}^\delta$ are the restrictions to the 0 and 1 components of the zeroth map of the unit of j_p^δ. Of course, the equalities which involve J_p^δ and $J_{\otimes p}^\delta$ require p to be odd, but we can construct a precisely analogous diagram

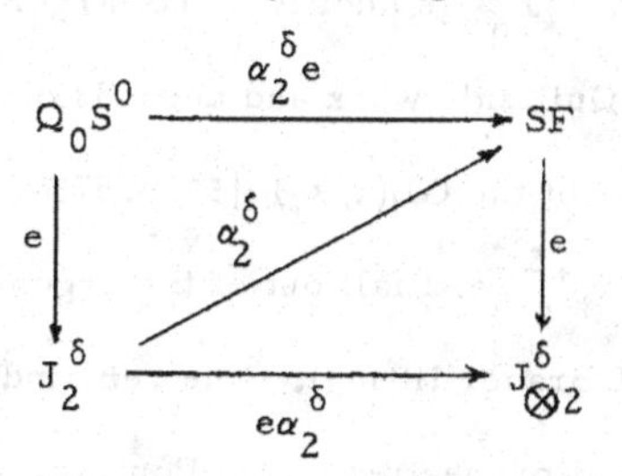

by use of $\mathcal{N}k_3$ rather than $\mathcal{B}\mathcal{L}k_r$. An analogous diagram can also be constructed by use of $\mathcal{O}k_r$. Henceforward, we assume that p is odd.

By the results of VII §4 and 5 cited above, all maps in our diagram are composites of maps of $\mathcal{Q}\times\mathcal{L}\times\mathcal{K}_\infty$-spaces and homotopy inverses of maps of $\mathcal{Q}\times\mathcal{L}\times\mathcal{K}_\infty$-spaces, and the diagram induces a similar commutative diagram in $H\mathcal{J}$. Clearly all three operads are required: it is $\mathcal{K}_\infty$ which acts naturally on Q_0S^0, $\mathcal{L}$ on SF, and $\mathcal{Q}$ on $B\mathcal{B}\mathcal{L}k_r$ (in two ways). Because of the different geometric sources of the actions, the statement that all of our maps preserve them is highly non-trivial.

<u>Theorem 4.1.</u> The composite $J_p^\delta \xrightarrow{\alpha_p^\delta} SF \xrightarrow{e} J_{\otimes p}^\delta$ is a homotopy equivalence.

<u>Proof.</u> By our diagram, the specified composite may also be described as

$$J_p^\delta = \Gamma_0(B\mathcal{B}\mathcal{L}k_r, \theta) \xrightarrow{\Gamma g} \Gamma_1(B_M\mathcal{B}\mathcal{L}k_r, \xi) \xrightarrow[\simeq]{\phi^{-1}} \Gamma_1(B\mathcal{B}\mathcal{L}k_r, \theta) = J_{\otimes p}^\delta.$$

It clearly suffices to prove that $\phi^{-1}\cdot\Gamma g$ induces an isomorphism on mod p homology H_*. By Quillen [59, §8], there exist elements y_i of degree $2i(p-1)$ and z_i of degree $2i(p-1)-1$, $i \geq 1$, such that

$$H_*J_p^\delta \cong H_*BGL(\infty, k_r) = P\{y_i\} \otimes E\{z_i\} .$$

Actually, we shall only need that $H_*J_p^\delta$ is additively no larger than stated. This is the easy part of Quillen's work and depends only on the form of the p-Sylow subgroups of the $GL(n; k_r)$ [59, p. 573-574]. The rest of the computation of $H_*J_p^\delta$ will fall out of the argument to follow (and is thus independent of Brauer lifting). The remainder of the proof depends solely on general properties of the Pontryagin products $*$ and # and the homology operations Q^s and $\tilde{Q}^s$ determined on E_∞ ring spaces by θ and ξ respectively [26, II §1, 2], together with particular properties of H_*QS^0 [26; I§4, II§5]. Write # by juxtaposition on elements, and write $[n]$ for the homology class of a component n. By [26, II.2.8], modulo linear combinations of $*$-products between positive degree elements,

(a) $\quad \tilde{Q}^s[r] \equiv \frac{1}{p}(r^p - r)Q^s[1] * [r^p - p]$.

The coefficient is non-zero because r reduces mod p^2 to a generator of the group of units of Z_{p^2}. By [26, II.2.8], for $x \in H_*Q_mS^0$ or $x \in H_*\Gamma_m(B\mathscr{B}\mathscr{U}k_r, \theta)$,

(b) $\quad mx \equiv (x*[1-m])[m]$.

Let $k = r^{-p}\frac{1}{p}(r^p - r)$; then (a) and (b) imply

(c) $\quad \tilde{Q}^s[r] \equiv k(Q^s[1]*[1-p])[r^p]$.

Since $e_r: QS^0 \to \Gamma(Q_M S^0, \xi)$ and Γg are exponential, they send

$$Q^s[1]*[-p] \quad \text{to} \quad \tilde{Q}^s[r][r^{-p}].$$

In view of (c), it follows that $(\Gamma i \circ \iota_\otimes)^{-1}e_r$ and $\phi^{-1}\Gamma g$ send

$$Q^s[1]*[-p] \quad \text{to} \quad k\,Q^s[1]*[1-p],$$

modulo elements decomposable under the translate $\underline{*}$ of the $*$-product from the zero component to the one component. By the multiplication table for # on H_*SF [26,II.5.6], it follows immediately that the composite

$$P\{Q^s[1]*[-p]\} \otimes E\{\beta Q^s[1]*[-p]\} \subset H_*Q_0S^0 \xrightarrow{(\Gamma i \circ \iota_{\otimes})_*^{-1}(e_r)_*} H_*SF$$

is a monomorphism (this being the step which would fail if $p = 2$). Since $(\Gamma i \circ \iota_{\otimes})^{-1} e_r$ factors through $e: Q_0S^0 \to J_p^\delta$, we conclude by a count of dimensions that

$$H_*J_p^\delta = P\{Q^q[1]*[-p]\} \otimes E\{\beta Q^s[1]*[-p]\}$$

as a Hopf algebra (under $*$) over the Steenrod algebra A. Moreover, by translation $x \to x*[1]$, we now have a basis in which we know $H_*J_{\otimes p}^\delta$ as a coalgebra over A (because J_p^δ and $J_{\otimes p}^\delta$ are the 0 and 1 components of $\Gamma B \mathcal{J} \not\chi k_r$). We already know that

$$\phi_*^{-1}(\Gamma g)_*(Q^1[1]*[-p]) = k\,Q^1[1]*[1-p] \neq 0,$$

and it follows by standard techniques that $\phi_*^{-1}(\Gamma g)_*$ is an isomorphism. In detail, $\phi_*^{-1}(\Gamma g)_*$ is a morphism of connected Hopf algebras of the same finite dimension in each degree and will be an isomorphism if it is a monomorphism on primitive elements. Let p_s, of degree $2s(p-1)$, be the sth even degree basic primitive element of $H_*J_p^\delta$, namely the sth Newton polynomial in the $Q^s[1]*[-p]$. Since, by [26,I.1.1], $P_*^r Q^s[1] = (-1)^r(r, s(p-1) - pr)Q^{s-r}[1]$, a standard calculation gives

$$P_*^r\, p_s = (-1)^r(r, s(p-1) - pr - 1)p_{s-r}\ .$$

Therefore some $P_*^r\, p_s \neq 0$, $r > 0$, unless $s = p^k$ for some $k \geq 0$ when

$$P_*^1\, p_{s+1} = p_s \quad \text{and} \quad P_*^p\, p_{s+1} = -p_{s+1-p} \quad \text{if} \quad k \geq 2$$

and

$$P_*^{p-1}\, p_{s+p-1} = p_s \quad \text{and} \quad P_*^p\, p_{s+p-1} = 2p_{s-1} \quad \text{if} \quad k = 1.$$

Thus, by induction on s, $\phi_*^{-1}(\Gamma g)_*(p_s) \neq 0$ for all $s \geq 1$. Let b_s, of degree $2s(p-1)-1$, be the sth odd degree basic primitive element of $H_* J_p^\delta$, so that $b_s \equiv \beta Q^s[1]*[-p]$ modulo elements decomposable under $*$ (and $\beta p_s = (-1)^{s+1} s b_s$). Since, again by [26, I.1.1], $P_*^r\, \beta Q^s[1] = (-1)^r(r, s(p-1) - pr - 1)\beta Q^{s-r}[1]$, another calculation gives

$$P_*^r\, b_s = (-1)^r(r, s(p-1) - pr - 1) b_{s-r}\, .$$

The coefficient here is the same as that in the even degree case, hence the same special cases show that $\phi_*^{-1}(\Gamma g)_*(b_s) \neq 0$ for all s since $\phi_*^{-1}(\Gamma g)_*(b_1) \neq 0$. The proof is complete.

In the following corollaries, we write $*$ or $\#$ for the product on infinite loop spaces according to whether we choose to think of them as additive or multiplicative. Recall from Corollary 3.3 that $(SF; j_p^\delta) = \Omega B(SF; j_p^\delta)$ is equivalent as an infinite loop space to $C_p^\delta = \Omega B C_p^\delta$.

<u>Corollary 4.2.</u> The composites

$$J_p^\delta \times (SF; j_p^\delta) \xrightarrow{\alpha_p^\delta \times \Omega q} SF \times SF \xrightarrow{\#} SF$$

and

$$BJ_p^\delta \times B(SF; j_p^\delta) \xrightarrow{B\alpha_p^\delta \times q} BSF \times BSF \xrightarrow{\#} BSF$$

are equivalences of infinite loop spaces.

<u>Proof.</u> Ωq is equivalent to the fibre of $e: SF \to J_{\otimes p}^\delta$, hence the theorem implies that the first composite, and thus also the second, induces an isomorphism on homotopy groups.

Choose an infinite loop map $w: BSF \to B(SF; j_p^\delta)$ such that $wq = 1$ as infinite loop maps.

Corollary 4.3. The map

$$B(SF;kO) \xrightarrow{(c(\psi^r), wq)} BO_{\otimes} \times B(SF;j_p^{\delta})$$

is an equivalence of infinite loop spaces.

Proof. Since $q: B(SF;j_p^{\delta}) \to BSF$ factors through $q: B(SF;kO^{\delta}) \to BSF$, this follows from V.4.4 (last line), V.4.8 (i), and the J-theory diagram of V§3 together with the multiplicative Brauer lift diagram of Theorem 3.4.

The original diagram of this section suggests that $(SF;j_p^{\delta}) \simeq C_p^{\delta}$ is the multiplicative analog of the additive infinite loop space $C_{\oplus p}^{\delta}$ defined as the fibre of $e: Q_0S^0 \to J_p^{\delta}$. Of course, we know that J_p^{δ} and $J_{\otimes p}^{\delta}$ are equivalent infinite loop spaces. In contrast, although $C_{\oplus p}^{\delta}$ and C_p^{δ} are evidently homotopy equivalent, there is no equivalence of infinite loop spaces between them because their homology operations differ [26;I §4, II §6].

Corollary 4.4. The composite

$$J_p^{\delta} \times C_{\oplus p}^{\delta} \xrightarrow{\tau^{-1}\alpha_p^{\delta} \times \pi} Q_0S^0 \times Q_0S^0 \xrightarrow{\ *\ } Q_0S^0$$

is a homotopy equivalence (but $\tau^{-1}\alpha_p^{\delta}$ is not an infinite loop map), where $\tau: Q_0S^0 \to SF$ is the translation $x \to x*1$.

In fact, by [26,I§4] and our proof of Theorem 4.1, the image of $(\tau^{-1}\alpha_p^{\delta})_*$ generates $H_*Q_0S^0$ as an algebra over the Dyer-Lashof algebra (under $*$ and the Q^s), and this statement even remains true at the prime 2. By [26, II§1], we also have the following technical consequence of our proof which has been used in the homological study of BF and BTop in [26,II]. Conceptually, we have here used crude information on homology operations to obtain the geometric splitting of SF, and we there used the geometric splitting to obtain more subtle information.

Corollary 4.5. $(\alpha_p^\delta)_*: H_*J_p^\delta \to H_*SF$ takes the elements $Q^s[1]*[-p]$ and $\beta Q^s[1]*[-p]$ to generators of the subalgebra

$$P\{Q^s[1]*[1-p]\} \otimes E\{\beta Q^s[1]*[1-p]\}$$

of H_*SF considered as an algebra under the $\underline{*}$ product.

The point is that no higher operations $Q^I[1]$, $\hat{\ell}(I) > 1$, contribute to the image of $(\alpha_p^\delta)_*$ on the specified generators. Since $(\alpha_p^\delta)_*$ is multiplicative with respect to $\#$, rather than $\underline{*}$, on H_*SF, such operations can contribute to the image of $(\alpha_p^\delta)_*$ on decomposable elements.

Remarks 4.6. The second author's original proof of Theorem 4.1 gave different information. Since p is odd, we may think of J_p and $J_{\otimes p}$ as the fibres of ψ^r-1 and $\psi^r/1$ on BU and $BU_\otimes$. Since $[J_p, U] = KU^{-1}BGL(\infty, k_r) = 0$ (by Theorem 3.2 and [14]), the composite equivalence $\varepsilon_p\alpha_p$ of V.4.6 is characterized by homotopy commutativity of the diagram

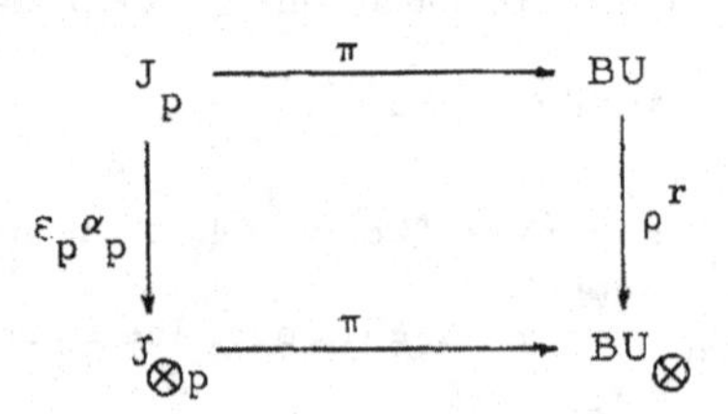

where ρ^r is the cannibalistic class determined by ψ^r and the standard orientation $BU \to B(U; kU)$. A representation theoretical calculation ([77, 4.1]) shows that the diagram

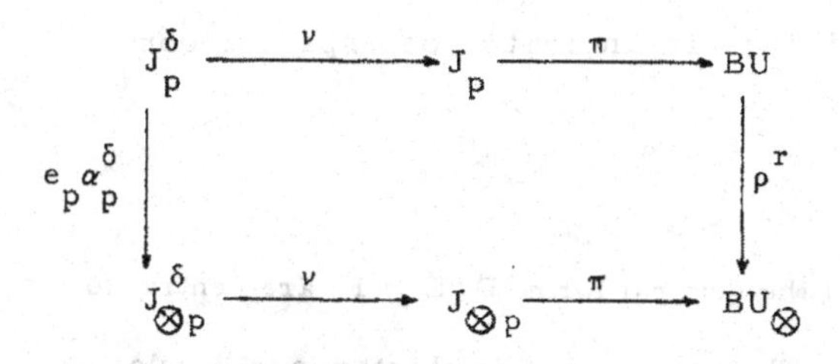

is homotopy commutative, where the maps ν are the 0 and 1 components of the zeroth map of the equivalence $\nu: j_p^\delta \to j_p$ of Theorem 3.2. Thus $\varepsilon_p\alpha_p \simeq \nu e\alpha_p^\delta\nu^{-1}$ and $e\alpha_p^\delta$ is an equivalence.

$\alpha_p^\delta\nu^{-1}: J_p \to SF$ is an infinite loop map while $\alpha_p: J_p \to SF$ makes the J-theory diagram homotopy commutative and, in particular, is such that $j: Spin \to SF$ factors through it. In view of Corollary 3.5 and the previous remarks, it is natural to hope that these two maps are homotopic or, at least, that α_p can be chosen as an infinite loop map. This would certainly hold if V.7.14 (the complex Adams conjecture on the infinite loop level) were satisfied.

Very recently, Friedlander [Stable Adams' conjecture. Preprint] and Seymour have announced proofs of Conjecture V.7.14. Unfortunately, Segal's machinery [68] seems essential to Friedlander's proof, hence it is not yet known that his infinite loop structure on SF agrees with ours (and ours is essential to such basic facets of the theory as the orientation sequences). We have not seen the details of Seymour's argument.

IX. Pairings in infinite loop space theory

Maps of spectra of the general form $D \wedge E \to F$ are central to stable homotopy theory. The purpose of this chapter is to develop a theory of pairings that allows one to recognize such maps in the guise of appropriate space level maps $X \wedge Y \to Z$, where X, Y, and Z are E_∞ spaces. Since $X \wedge Y$ will not itself be an E_∞ space, such a theory is certainly not implicit in the recognition principle already obtained in chapter VII. It will be convenient to work with (weak) prespectra and their pairings in the sense of Whitehead [80] throughout this chapter. The relationship between these notions and the stable category has been explained in II §3.

While a theory of pairings is an obvious desideratum of any complete treatment of infinite loop space theory and should have many other applications, the need for it emerged in attempts to compare our machine-built spectra $\Omega^\infty TB\mathscr{P}A$, where $\mathscr{P}A$ is the category of finitely generated projective modules over a ring A, to the Gersten-Wagoner spectra [30, 79]. Let CA be the ring of infinite, but row and column finite, matrices with entries in A and let SA be the quotient of CA by the ideal generated by the finite matrices. Gersten and Wagoner showed that ΩKSA is equivalent to KA, where KA denotes $BGL(\infty, A)^+ \times K_0A$, and thus produced an Ω-prespectrum $GWA = \{KS^iA \mid i \geq 0\}$.

Since free modules are cofinal among projective modules, $BGL(\infty, A)$ may be regarded as $\varinjlim B\,\mathrm{Aut}\,P$, $P \in \mathscr{P}A$ (up to homotopy type; compare [46, p. 85]). By the universal property of the plus construction (above VIII.1.1), the tensor product functor $\mathscr{P}A \times \mathscr{P}B \to \mathscr{P}(A \otimes_Z B)$ induces a map

$\mu: KA \wedge KB \rightarrow K(A \otimes_Z B)$. Suppose given a functor E from rings to Ω-prespectra, written $EA = \{E_i A \mid i \geq 0\}$, such that $E_0 A = KA$ and suppose that E admits an external tensor product, by which we understand a natural pairing $(EA, EB) \rightarrow E(A \otimes_Z B)$ which extends the map μ of zeroth spaces. With these data, Fiedorowicz [27] has proven that there is a natural map $f: EA \rightarrow GWA$ of Ω-prespectra such that $f_0: E_0 A = KA \rightarrow KA$ is the identity. It follows (by II. 2.10 and 2.11) that the associated connective spectra of EA and GWA are equivalent.

Now let EA denote $\Omega^\infty TB\mathcal{P}A$ regarded as an Ω-prespectrum. Certainly $E_0 A$ is KA (up to homotopy type). The results of this chapter will imply that E admits an external tensor product and thus that EA is the associated connective spectrum of GWA.

We develop suitably related notions of pairings of symmetric monoidal categories, of permutative categories, and of $\mathcal{Q}$-spaces in section 1. We prove that pairings of $\mathcal{Q}$-spaces induce pairings of Ω-prespectra in section 2. Schematically, our results can be summarized as follows:

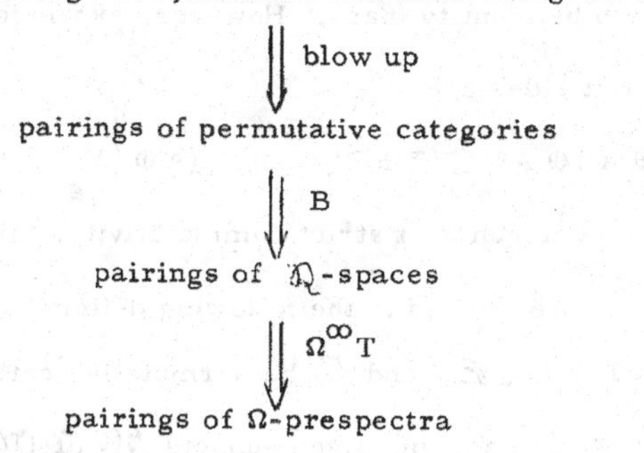

While there is an evident intuitive relationship between the present theory and the E_∞ ring theory of the earlier chapters, I have not attempted to combine the two lines of thought into a single more general theory.

1. Pairings of categories and of $\mathcal{Q}$-spaces

Let $\mathcal{A}$, $\mathcal{B}$, and $\mathcal{C}$ be symmetric monoidal categories with products $\oplus$ and units 0. A pairing $\otimes : \mathcal{A} \times \mathcal{B} \to \mathcal{C}$ is a functor $\otimes$ such that $A \otimes 0 = 0$ and $0 \otimes B = 0$ together with a coherent natural bidistributivity isomorphism

$$(*)\qquad d: (A \oplus A') \otimes (B \oplus B') \cong ((A \otimes B) \oplus (A \otimes B')) \oplus ((A' \otimes B) \oplus (A' \otimes B'))$$

for $A, A' \in \mathcal{O}\mathcal{A}$ and $B, B' \in \mathcal{O}\mathcal{B}$; the extra parentheses are needed since $\oplus$ on $\mathcal{C}$ is not assumed to be associative.

The category theorist will recognize that this is not really a definition. Precision would require elucidation of the meaning of coherence, via a specification of just which diagrams involving d and the associativity, unity, and commutativity isomorphisms a, b, and c are required to commute. The details would be analogous to those in La Plaza [35]. We prefer to be informal since the intuition should be clear. Of course, the example to keep in mind is $\otimes : \mathcal{F}A \times \mathcal{F}B \to \mathcal{P}(A \otimes_Z B)$.

We would like to define a pairing of permutative categories by requiring the isomorphisms d to be identity maps. However, expansion of the right sides of (*) when the left sides are

$$((A \oplus A') \oplus A'') \otimes (B \oplus B') \quad \text{and} \quad (A \oplus (A' \oplus A'')) \otimes (B \oplus B')$$

demonstrates that, in the absence of strict commutativity, this requirement would be unreasonable. We are led to the following definition.

Definition 1.1. Let $\mathcal{A}$, $\mathcal{B}$, and $\mathcal{C}$ be permutative categories, with products $\oplus$ and units 0, and assume given subsets $\mathcal{G}\mathcal{A}$ of $\mathcal{O}\mathcal{A}$ and $\mathcal{G}\mathcal{B}$ of $\mathcal{O}\mathcal{B}$ which generate $\mathcal{O}\mathcal{A}$ and $\mathcal{O}\mathcal{B}$ under $\oplus$. A pairing $\otimes : \mathcal{A} \times \mathcal{B} \to \mathcal{C}$ is a functor $\otimes$ such that $A \otimes 0 = 0$, $0 \otimes B = 0$, and for all sequences $\{A_1, \ldots, A_j\}$ of objects in $\mathcal{G}\mathcal{A}$ and all sequences $\{B_1, \ldots, B_k\}$ of objects in $\mathcal{G}\mathcal{B}$

$$(**)\quad (A_1\oplus\ldots\oplus A_j)\otimes(B_1\oplus\ldots\oplus B_k) = (A_1\otimes B_1)\otimes\ldots\otimes(A_1\otimes B_k)\otimes\ldots\otimes (A_j\otimes B_1)\otimes\ldots\otimes(A_j\otimes B_k)$$

and the following diagram commutes for all permutations $\sigma\in\Sigma_j$ and $\tau\in\Sigma_k$, where the unlabelled isomorphisms are given by the commutativity isomorphisms of $\mathcal{A}$, $\mathcal{B}$, and $\mathcal{C}$:

$$\begin{array}{ccc} (A_1\oplus\ldots\oplus A_j)\otimes(B_1\oplus\ldots\oplus B_k) & = & (A_1\otimes B_1)\oplus\ldots\oplus(A_j\otimes A_k) \\ \wr\| & & \wr\| \\ (A_{\sigma(1)}\oplus\ldots\oplus A_{\sigma(j)})\otimes(B_{\tau(1)}\oplus\ldots\oplus B_{\tau(k)}) & = & (A_{\sigma(1)}\otimes B_{\tau(1)})\oplus\ldots\oplus(A_{\sigma(j)}\otimes A_{\tau(k)}) \end{array}$$

By use of the commutativity isomorphism of $\mathcal{C}$, (**) determines a natural bidistributivity isomorphism d as in (*). The commutative diagrams above give coherence. Thus a pairing of permutative categories is also a pairing of symmetric monoidal categories. It is an instructive exercise to verify that $\otimes\colon \mathcal{GL}A\times\mathcal{GL}B\to\mathcal{GL}(A\otimes_Z B)$ is a pairing of permutative categories. In this case, the generating sets of objects have the single element 1, (**) is trivial, and only the diagrammatic relationship between the commutativity isomorphisms need be checked.

Recall the functor Φ of VI.3.2 from symmetric monoidal categories to permutative categories.

<u>Proposition 1.2.</u> A pairing $\otimes\colon\mathcal{A}\times\mathcal{B}\to\mathcal{C}$ of symmetric monoidal categories naturally determines a pairing $\otimes\colon\Phi\mathcal{A}\times\Phi\mathcal{B}\to\Phi\mathcal{C}$ of permutative categories such that the diagram

$$\begin{array}{ccc} \Phi\mathcal{A}\times\Phi\mathcal{B} & \xrightarrow{\otimes} & \Phi\mathcal{C} \\ \downarrow{\scriptstyle \pi\times\pi} & & \downarrow{\scriptstyle \pi} \\ \mathcal{A}\times\mathcal{B} & \xrightarrow{\otimes} & \mathcal{C} \end{array}$$

commutes up to coherent natural isomorphism.

Proof. The space of objects of $\Phi\mathcal{A}$ is the free monoid with unit 0 generated by $\mathcal{O}\mathcal{A}$, hence we take $\mathcal{O}\mathcal{A}$ as the generating set in $\mathcal{O}\Phi\mathcal{A}$ and similarly for $\mathcal{B}$. Recall that, with the product on $\Phi\mathcal{A}$ again written as $\oplus$, π is specified on objects by

$$\pi(A_1 \oplus \ldots \oplus A_j) = A_1 \oplus (A_2 \oplus (A_3 \oplus \ldots (A_{j-1} \oplus A_j) \ldots)) .$$

As in [46, 4.2] or VI.3.5, the morphisms from A to A' in $\Phi\mathcal{A}$ are the morphisms from πA to $\pi A'$ in $\mathcal{A}$, with composition and the commutativity isomorphism c determined in an evident way from these data on $\mathcal{A}$. Define $\otimes: \Phi\mathcal{A} \times \Phi\mathcal{B} \to \Phi\mathcal{C}$ by (**) on objects. On morphisms $f: A \to A'$ and $g: B \to B'$ in $\Phi\mathcal{A}$ and $\Phi\mathcal{B}$, the morphism $f \otimes g: A \otimes B \to A' \otimes B'$ in $\Phi\mathcal{C}$ is specified by the composite

$$\pi(A \otimes B) \cong \pi(A) \otimes \pi(B) \xrightarrow{f \otimes g} \pi(A') \otimes \pi(B') \cong \pi(A' \otimes B')$$

in $\mathcal{C}$, where the unlabelled isomorphisms are uniquely determined by the coherent natural isomorphisms a, c, and d of $\mathcal{C}$ and are the isomorphisms required for the diagram in the statement of the proposition. The commutativity of the diagram in Definition 1.1 follows from the coherence of the given pairing $\otimes: \mathcal{A} \times \mathcal{B} \to \mathcal{C}$. Indeed, the omitted formal definition of coherence here can be specified simply by listing those diagrams which suffice for the present proof.

Now recall the categorical E_∞ operad $\mathcal{Q}$ of VI§4. The tensor product $\Sigma_j \times \Sigma_k \to \Sigma_{jk}$ (of Notations VI.1.4) induces a functor $\tilde{\Sigma}_j \times \tilde{\Sigma}_k \to \tilde{\Sigma}_{jk}$ and thus, by application of the classifying space functor B, a map $\otimes: \mathcal{Q}(j) \times \mathcal{Q}(k) \to \mathcal{Q}(jk)$.

Definition 1.3. Let X, Y, and Z be $\mathcal{Q}$-spaces. A pairing $f: X \times Y \to Z$ is a map f which factors through $X \wedge Y$ and is such that the following diagram commutes:

$$\begin{array}{ccc}
\mathcal{Q}(j)\times X^j\times\mathcal{Q}(k)\times Y^k & \xrightarrow{\theta_j\times\theta_k} & X\times Y \\
\downarrow{\scriptstyle 1\times\tau\times 1} & & \searrow{\scriptstyle f} \\
\mathcal{Q}(j)\times\mathcal{Q}(k)\times X^j\times Y^k & & Z \\
\downarrow{\scriptstyle 1\times 1\times\nu} & & \nearrow{\scriptstyle \theta_{jk}} \\
\mathcal{Q}(j)\times\mathcal{Q}(k)\times(X\wedge Y)^{jk} & \xrightarrow{\otimes\times f^{jk}} & \mathcal{Q}(jk)\times Z^{jk}
\end{array}$$

where

$$\nu(x_1,\ldots,x_j,y_1,\ldots,y_k) = (x_1\wedge y_1,\ldots,x_1\wedge y_k,\ldots,x_j\wedge y_1,\ldots,x_j\wedge y_k)\ .$$

I do not have a definition (or any prospective applications) for a notion of pairing of $\mathcal{C}$-spaces for a general E_∞ operad $\mathcal{C}$. One could, of course, simply appeal to VI. 2. 7 (iii), which shows that $\mathcal{C}$-spaces can be replaced by equivalent $\mathcal{Q}$-spaces.

<u>Proposition 1.4.</u> If $\otimes: \mathcal{A}\times\mathcal{B}\to\mathcal{C}$ is a pairing of permutative categories, then $f = B\otimes: B\mathcal{A}\times B\mathcal{B}\to B\mathcal{C}$ is a pairing of $\mathcal{Q}$-spaces.

<u>Proof.</u> The basepoint of $B\mathcal{A}$ is given by the object 0 (regarded as a 0-simplex), and f factors through $B\mathcal{A}\wedge B\mathcal{B}$ by the nullity of zero. The diagram of Definition 1.1 implies the commutativity of the following coherence diagram:

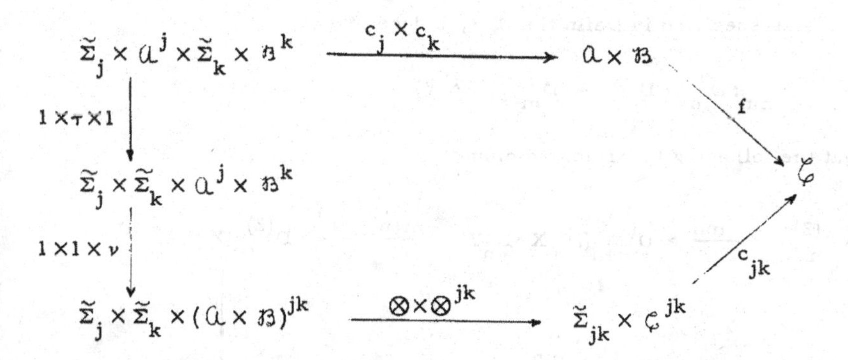

where the functor ν is defined just as was the map ν in the previous definition. The conclusion follows upon application of B.

2. The recognition principle for pairings

We here extend the one operad recognition principle of VII §3 to a recognition principle for pairings. Although the present theory is basically an elaboration of the additive theory, it will still be necessary, for technical reasons, to work with the little convex bodies (partial) operads $\mathcal{K}_n$ rather than with the little cubes operads $\mathcal{C}_n$. Defin $\otimes: \mathcal{K}_m(j) \times \mathcal{K}_n(k) \to \mathcal{K}_{m+n}(jk)$ by

$$\langle c_1,\ldots,c_j\rangle \otimes \langle c_1',\ldots,c_k'\rangle = \langle c_1 \times c_1',\ldots,c_1 \times c_k',\ldots,c_j \times c_1',\ldots,c_j \times c_k'\rangle .$$

Let $\mathcal{Q}_n$ denote $\mathcal{Q} \times \mathcal{K}_n$ for $n \geq 1$. The maps $\otimes: \mathcal{Q}(j) \times \mathcal{Q}(k) \to \mathcal{Q}(jk)$ and the maps just defined together determine maps $\otimes: \mathcal{Q}_m(j) \times \mathcal{Q}_n(k) \to \mathcal{Q}_{m+n}(jk)$. Let D_n denote the (partial) monad associated to $\mathcal{Q}_n$. We begin by using the maps $\otimes$ to define a "pairing of monads" $D_m \wedge D_n \to D_{m+n}$.

Proposition 2.1. For based spaces X and Y, the composite maps

$$\mathcal{Q}_m(j) \times X^j \times \mathcal{Q}_n(k) \times Y^k \xrightarrow{1 \times \tau \times 1} \mathcal{Q}_m(j) \times \mathcal{Q}_n(k) \times X^j \times Y^k \xrightarrow{\otimes \times \nu} \mathcal{Q}_{m+n}(jk) \times (X \wedge Y)^{jk} ,$$

where ν is as specified in Definition 1.3, induce maps

$$\lambda_{mn}: D_m X \wedge D_n Y \to D_{m+n}(X \wedge Y)$$

such that the following diagrams commute:

$$\begin{array}{ccccc} D_m^{(2)} X \wedge D_n^{(2)} Y & \xrightarrow{\lambda_{mn}} & D_{m+n}^{(1)}(D_m X \wedge D_n Y) & \xrightarrow{D_{m+n}(\lambda_{mn})} & D_{m+n}^{(2)}(X \wedge Y) \\ \downarrow{\scriptstyle \mu \wedge \mu} & & & & \downarrow{\scriptstyle \mu} \\ D_m X \wedge D_n Y & & \xrightarrow{\lambda_{mn}} & & D_{m+n}(X \wedge Y) \end{array}$$

and

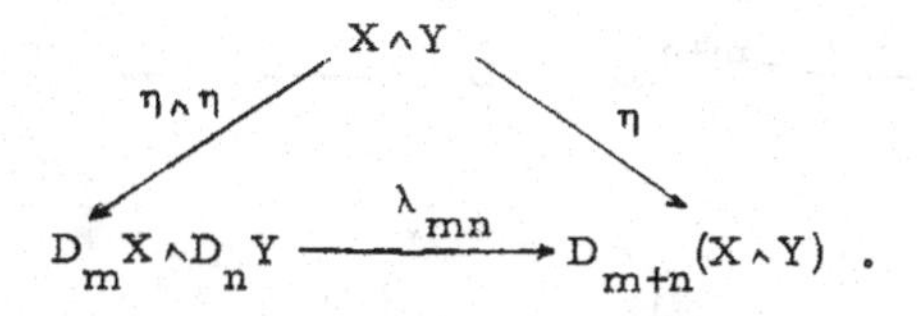

Proof. $D_m X$ is constructed from $\coprod \mathcal{Q}_m(j) \times X^j$ by use of appropriate equivariance and basepoint identifications, and its product μ and unit η are induced from the structural maps γ and unit 1 for the operads $\mathcal{Q}$ and $\mathcal{K}_m$ (see VI. 1.2 and [45, 2. 4]) . The proof consists of a check, in principle for $\mathcal{Q}$ and the $\mathcal{K}_m$ separately, of the commutation relations between $\otimes$ and the defining data of the specified operads. The details are closely analogous to those already formulated in VI. 1. 6-1. 10 (specialized to the case k = 2) and will therefore be omitted. The top row of the first diagram must be interpreted in the sense of partial monads, the superscripts indicating restrictions of powers such that only composable pairs of little convex bodies are in sight (see VII. 1.1 and the discussion following VII. 2.1). For $D^{(1)}_{m+n}(D_m X \wedge D_n Y)$, composability is to be interpreted in terms of $\otimes$ on the $\mathcal{K}$'s. Indeed, we may specify this space to be the inverse image of $D^{(2)}_{m+n}(X \wedge Y)$ under $D_{m+n}(\lambda_{mn})$ and then check that it contains the image of $D^{(2)}_m X \wedge D^{(2)}_n Y$ under the map

$$\lambda_{mn}: D_m D_m X \wedge D_n D_n Y \rightarrow D_{m+n}(D_m X \wedge D_n Y).$$

Let $\beta_n : \Sigma^n D_n \rightarrow D_n$ be the adjoint of the composite of the projection $\pi: D_n \rightarrow K_n$ and the morphism of monads $\alpha_n : K_n \rightarrow \Omega^n \Sigma^n$ of VII. 2.2. Recall (for twisting maps) that we are writing suspension coordinates on the right.

Proposition 2.2. The following diagrams commute for all X and Y

$$\begin{array}{ccccc} \Sigma^m D_m X \wedge \Sigma^n D_n Y & \xrightarrow{1\wedge T\wedge 1} & \Sigma^{m+n}(D_m X \wedge D_n Y) & \xrightarrow{\Sigma^{m+n}\lambda_{mn}} & \Sigma^{m+n} D_{m+n}(X\wedge Y) \\ \downarrow \beta_m \wedge \beta_n & & & & \downarrow \beta_{m+n} \\ \Sigma^m X \wedge \Sigma^n Y & & \xrightarrow{1\wedge T\wedge 1} & & \Sigma^{m+n}(X\wedge Y) \end{array}$$

Proof. We may define $\lambda_{mn}: K_m X \wedge K_n Y \to K_{m+n}(X\wedge Y)$ just as in the previous proposition, and then $\pi\lambda_{mn} = \lambda_{mn}(\pi\wedge\pi)$. Moreover, as was pointed out in [45, 8.3] in the case of little cubes, the following diagram is commutative:

$$\begin{array}{ccc} K_m X \wedge K_n Y & \xrightarrow{\lambda_{mn}} & K_{m+n}(X\wedge Y) \\ \downarrow \alpha_m \wedge \alpha_n & & \downarrow \alpha_{m+n} \\ \Omega^m \Sigma^m X \wedge \Omega^n \Sigma^n Y & \xrightarrow{\wedge} & \Omega^{m+n}\Sigma^{m+n}(X\wedge Y) \end{array}$$

The conclusion follows by passage to adjoints.

Recall that $\sigma: \mathcal{K}_n \to \mathcal{K}_{n+1}$ is the morphism of operads specified by $c \to c \times 1$ on little convex bodies, and also write σ for $1 \times \sigma: \mathcal{K}'_n \to \mathcal{K}'_{n+1}$. We need to know that, up to homotopy, σ is independent of the choice of privileged coordinate. The following analog of [45, 4.9] for little cubes will give the idea. Let $\sigma': \mathcal{K}_n \to \mathcal{K}_{n+1}$ be specified by $c \to 1 \times c$ on little convex bodies.

Lemma 2.3. The maps σ and σ' from $\mathcal{K}_n(j)$ to $\mathcal{K}_{n+1}(j)$ are Σ_j-equivariantly homotopic.

Proof. Define orthogonal transformations $g, g': R^{n+1} \to R^{n+1}$ by

$$g(s,x) = (x,s) \quad\text{and}\quad g'(s,x) = \begin{cases} (s,x) & \text{if } n \text{ is even} \\ (1-s,x) & \text{if } n \text{ is odd} \end{cases}$$

for $x \in R^n$ and $s \in R$. Since g and g' both have degree $(-1)^n$, there is a path $h: I \to O(n)$ from g to g'. For a little convex body $c: R^n \to R^n$, we have

$$g\sigma'(c)g^{-1} = \sigma(c) \quad\text{and}\quad g'\sigma'(c)(g')^{-1} = \sigma'(c).$$

The required homotopy is given on little convex bodies c by conjugation of $\sigma'(c)$ with the orthogonal transformations h_t.

It is not just the existence but the form of the homotopies that is essential for our purposes. For example, the following result is immediate from the previous proof.

Proposition 2.4. The bottom part of the following diagram commutes and the top part commutes up to homotopy for all X and Y:

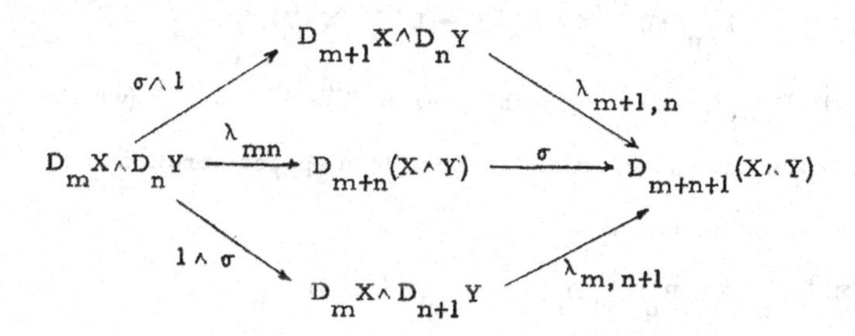

At this point, we recall Whitehead's definition [80] of a pairing.

Definition 2.5. A pairing $\phi: (T', T'') \rightarrow T$ of prespectra consists of maps $\phi_{mn}: T'_m \wedge T''_n \rightarrow T_{m+n}$ for $m, n \geq 0$ such that, up to homotopy, the bottom part of the following diagram commutes and the top part of the diagram commutes up to the sign $(-1)^n$:

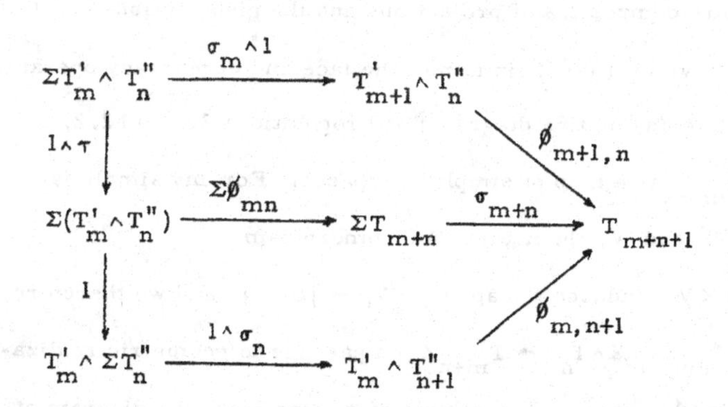

For a $\mathcal{R}$-space X, we have the prespectrum TX specified by

$$T_iX = B(\Sigma^i, D_i, X),$$

where $T_0X = X$, with structural maps

$$\sigma_i = B(1,\sigma,1): \Sigma T_i X \cong B(\Sigma^{i+1}, D_i, X) \to B(\Sigma^{i+1}, D_{i+1}, X) = T_{i+1}X .$$

We have the following recognition principle for pairings.

Theorem 2.6. A pairing $f: X \wedge Y \to Z$ of $\mathcal{Q}$-spaces naturally induces a pairing $\phi: (TX, TY) \to TZ$ of prespectra such that $\phi_{0,0} = f$.

Proof. As in the first diagram of Proposition 2.1, the maps λ_{mn} can be iterated to yield

$$\lambda_{mn}^{(q)}: D_m^{(q)} X \wedge D_n^{(q)} Y \to D_{m+n}^{(q)}(X \wedge Y).$$

By composing with $D_{m+n}^{(q)} f$, smashing with spheres $S^m \wedge S^n = S^{m+n}$ (which are taken as one-point compactifications of Euclidean spaces here), and using a twist map τ, we obtain maps

$$\phi_{mnq}: B_q(\Sigma^m, D_m, X) \wedge B_q(\Sigma^n, D_n, Y) \to B_q(\Sigma^{m+n}, D_{m+n}, Z) .$$

By the definition of a pairing of $\mathcal{Q}$-spaces, we have the commutative diagrams

$$\begin{array}{ccccc} D_m X \wedge D_n Y & \xrightarrow{\lambda_{mn}} & D_{m+n}(X \wedge Y) & \xrightarrow{D_{m+n} f} & D_{m+n} Z \\ \downarrow{\scriptstyle \theta \wedge \theta} & & & & \downarrow{\scriptstyle \theta} \\ X \wedge Y & & \xrightarrow{\quad\quad\quad} & & Z \end{array}$$

where the θ are composites of projections and the given actions of D on X, Y, and Z. In view of the definition of the face and degeneracy operators [45, 9.6] and the commutative diagrams of Propositions 2.1 and 2.2, it follows that ϕ_{mn*} is a map of simplicial spaces. For any simplicial based spaces U and V, the natural homeomorphism $|U| \times |V| \cong |U \times V|$ induces a map $|U| \wedge |V| \to |U \wedge V|$, and we therefore obtain a map $\phi_{mn}: T_m X \wedge T_n Y \to T_{m+n} Z$ on passage to geometric realization. Certainly $\phi_{0,0} = f$. On the level of q-simplices, the diagram of the previous definition can be written as follows:

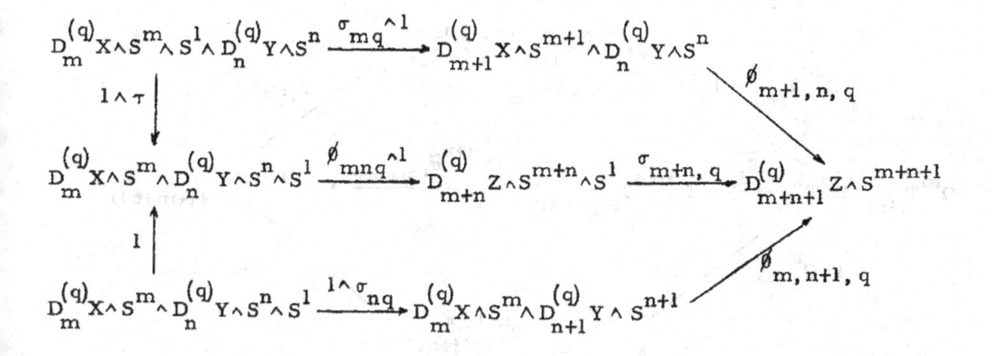

The bottom part commutes by Proposition 2.4. Provided that we first twist S^n past S^1, application of the homotopy of Proposition 2.4 to $D_m^{(q)}X \wedge D_n^{(q)}Y$ and of the orthogonal transformations which give that homotopy to S^{m+1+n} yields a homotopy for the top part of the diagram. These homotopies as q varies are compatible with the face and degeneracy operators (for each parameter $t \in I$) and so determine the required homotopy on passage to geometric realization. It is for this compatibility with face operators that use of little convex bodies rather than little cubes is essential.

While the passage via Ω^∞ from pairings of prespectra to maps in the stable category has already been discussed in II. 3.3 and 3.4, we should perhaps say a bit about the more elementary passage from pairings of prespectra to pairings of Ω-prespectra. Provided that we are willing to neglect phantom maps, the functor Ω^∞ can be redefined homotopically by

$$(\Omega^\infty T)_i = \mathrm{Tel}\, \Omega^j T_{i+j},$$

with $\iota : T_i \to (\Omega^\infty T)_i$ being given by the 0th term of the limit system. Given a pairing $\phi: (T', T'') \to T$, the maps

$$\bar{\phi}_{ij}: \Omega^m T'_{i+m} \wedge \Omega^n T''_{j+n} \to \Omega^{m+n} T_{i+j+m+n}$$

specified by $\bar{\phi}_{ij}(f \wedge g)(x \wedge y) = \phi_{i+m,\, j+n}(f(x) \wedge g(y))$ for $x \in S^m$ and $y \in S^n$ are such that the following diagrams are homotopy commutative:

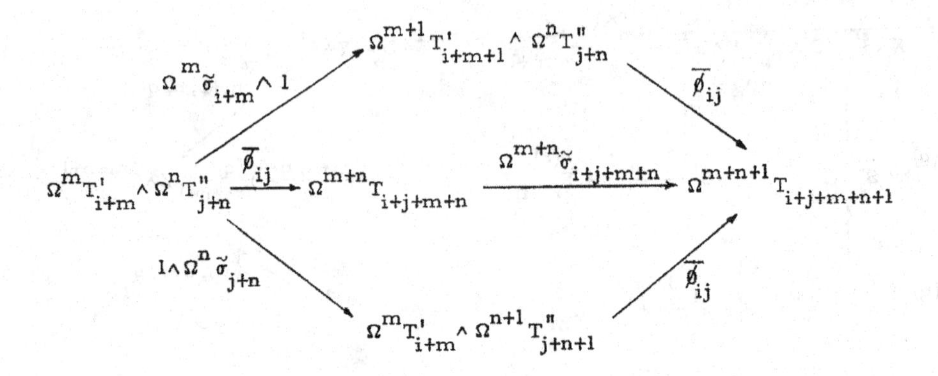

there being a permutation of loop coordinates in the upper part which cancels the sign inserted in the definition of a pairing. Still neglecting phantom maps, there result maps

$$\tilde{\phi}_{ij}: (\Omega^{\infty}T')_i \wedge (\Omega^{\infty}T'')_j \simeq \mathrm{Tel}(\Omega^m T'_{i+m} \wedge \Omega^n T''_{j+n}) \rightarrow (\Omega^{\infty}T)_{i+j}$$

which give a pairing $\tilde{\phi}: (\Omega^{\infty}T', \Omega^{\infty}T'') \rightarrow \Omega^{\infty}T$. The following diagrams are clearly homotopy commutative:

$$\begin{array}{ccc} T'_i \wedge T''_j & \xrightarrow{\phi_{ij}} & T \\ \downarrow{\iota \wedge \iota} & & \downarrow{\iota} \\ (\Omega^{\infty}T')_i \wedge (\Omega^{\infty}T'')_j & \xrightarrow{\tilde{\phi}_{ij}} & (\Omega^{\infty}T)_{i+j} \end{array}$$

In the context of Theorem 2.5, the group completion property of the recognition principle implies that the map ϕ_{00} is characterized by the case $i = j = 0$ of this diagram (compare [46, 3.9], VII. 1.1, and the paragraphs above the latter result).

One could obtain a genuine pairing $\tilde{\phi}$, without neglect of phantom maps, by an elaboration of the discussion just given in terms of the mapping cylinder techniques of [43, Theorem 4]. However, the extra precision would be insignificant in view of II. 3. 4.

Bibliography

1. J.F. Adams. Vector fields on spheres. Annals of Math. 75(1962), 603-632

2. J.F. Adams. On the groups J(X)-I. Topology 2(1963), 181-195.

3. J.F. Adams. On the groups J(X)-II. Topology 3 (1965), 137-171.

4. J.F. Adams. On the groups J(X)-III. Topology 3(1965), 193-222.

5. J.F. Adams. On the groups J(X)-IV. Topology 5(1966), 21-71.

6. J.F. Adams. Lectures on generalized homology. Lecture Notes in Mathematics, Vol. 99, 1-138. Springer, 1969.

7. J.F. Adams. Stable homotopy and generalized homology. The University of Chicago Press, 1974.

8. J.F. Adams and S. Priddy. Uniqueness of BSO. Preprint.

9. D.W. Anderson. Thesis. Berkeley, 1963.

10. D.W. Anderson. Simplicial K-theory and generalized homology theories. Preprint.

11. D.W. Anderson. There are no phantom cohomology operations in K-theory. Preprint.

12. D.W. Anderson and L. Hodgkin. The K-theory of Eilenberg-MacLane complexes. Topology 7(1968), 317-329.

13. M.F. Atiyah, R. Bott, and A. Shapiro. Clifford modules. Topology 3 Supp. 1 (1964), 3-38.

14. M.F. Atiyah and G.B. Segal. Equivariant K-theory and completion. J. Diff. Geometry 3(1969), 1-18.

15. M.F. Atiyah and D.O. Tall. Group representations, λ-rings and the J-homomorphism. Topology 8 (1969), 253-297.

16. M. G. Barratt and P. J. Eccles. Γ^+-structures - I, II. Topology 13(1974), 25-45 and 113-126.

17. J. C. Becker and D. H. Gottlieb. The transfer map and fiber bundles. Topology 14(1975), 1-14.

18. J. M. Boardman. Stable homotopy theory. Mimeographed notes.

19. J. M. Boardman and R. M. Vogt. Homotopy everything H-spaces. Bull. Amer. Math. Soc. 74 (1968), 1117-1122.

20. J. M. Boardman and R. M. Vogt. Homotopy invariant algebraic structures on topological spaces. Lecture Notes in Mathematics, Vol. 347. Springer, 1973.

21. R. Bott. Quelques remarques sur les théorèmes de périodicité. Bull. Soc. Math. France 87 (1959), 293-310.

22. R. Bott. A note on the KO-theory of sphere bundles. Bull. Amer. Math. Soc. 68 (1962), 395-400.

23. A. K. Bousfield and D. M. Kan. Homotopy limits, completions, and localizations. Lecture Notes in Mathematics, Vol. 304 Springer, 1972.

24. G. Brumfiel. On the homotopy groups of BPL and PL/O. Annals Math. 88(1968), 291-311.

25. H. Cartan et al. Périodicité des groupes d'homotopie stables des groupes classiques, d'après Bott. Séminaire Henri Cartan, 1959/60.

26. F. Cohen, T. Lada, and J. P. May. The Homology of Iterated Loop Spaces. Lecture Notes in Mathematics, Vol. 533. Springer, 1976.

27. Z. Fiedorowicz. A note on the spectra of algebraic K-theory. Preprint.

28. Z. Fiedorowicz and S. Priddy. Loop spaces and finite orthogonal groups. Bull. Amer. Math. Soc. 81(1975), 700-702.

29. E. Friedlander. Computations of K-theories of finite fields. Topology 15(1976), 87-109.

30. S. Gersten. On the spectrum of algebraic K-theory. Bull. Amer. Math. Soc. 78(1972), 216-219.

31. J. A. Green. The characters of the finite general linear groups. Trans. Amer. Math. Soc. 80(1955), 402-447.

32. A. Haefliger. Lissage des immersions - II. Preprint.

33. L. Hodgkin and V. P. Snaith. The K-theory of some more well-known spaces. Preprint.

34. M. Karoubi. Périodicité de la K-théorie Hermitienne. Lecture Notes in Mathematics, Vol. 343, 301-411. Springer, 1972.

35. M. Laplaza. Coherence for distributivity. Lecture Notes in Mathematics, Vol. 281, 29-65. Springer, 1972.

36. R. K. Lashof. Poincaré duality and cobordism. Trans. Amer. Math. Soc. 109(1963), 257-277.

37. R. Lee and R. H. Szczarba. On $K_3(Z)$. Preprint.

38. H. J. Ligaard. Infinite loop maps from SF to $BO_{\otimes}$ at the prime 2. Preprint.

39. H. J. Ligaard and I. Madsen. Homology operations in the Eilenberg-Moore spectral sequence. Math. Zeitschrift 143(1975), 45-54.

40. S. Mac Lane. Categories for the Working Mathematician. Springer, 1971.

41. I. Madsen. On the action of the Dyer-Lashof algebra in $H_*(G)$. Pacific J. Math. 60(1975), 235-275.

42. I. Madsen, V. P. Snaith, and J. Tornehave. Infinite loop maps in geometric topology. Preprint.

43. J. P. May. Categories of spectra and infinite loop spaces. Lecture Notes in Mathematics, Vol. 99, 448-479. Springer, 1969.

44. J.P. May. Homology operations on infinite loop spaces. Proc. Symp. Pure Math. Vol. 22, 171-186. Amer. Math. Soc. 1971.

45. J.P. May. The geometry of iterated loop spaces. Lecture Notes in Mathematics, Vol. 271. Springer, 1972.

46. J.P. May. E_∞ spaces, group completions, and permutative categories. London Math. Soc. Lecture Note Series 11, 1974, 61-94.

47. J.P. May. Classifying spaces and fibrations. Memoirs Amer. Math. Soc. No. 155, 1975.

48. J.P. May. The Homotopical Foundations of Algebraic Topology. Monograph London Math. Soc. Academic Press. In preparation.

49. J.P. May. Infinite loop space theory. Bull. Amer. Math. Soc. To appear.

50. J. Milnor and J.C. Moore. On the structure of Hopf algebras. Annals Math. 81(1965), 211-264.

51. O.T. O'Meara. Introduction to Quadratic Forms. Springer, 1963.

52. R.R. Patterson and R.E. Stong. Orientability of bundles. Duke Math. J. 39(1972), 619-622.

53. F.P. Peterson. Mod p homotopy type of BSO and F/PL. Bol. Soc. Mat. Mexicana 14(1969), 22-27.

54. S. Priddy. Dyer-Lashof operations for the classifying spaces of certain matrix groups. Quart. J. Math. Oxford (3) 36(1975), 179-193.

55. D. Puppe. Stabile Homotopietheorie I. Math. Annalen 169(1967), 243-274.

56. D. Puppe. On the stable homotopy category. Topology and its applications. Budva 1972 (Beograd 1973), 200-212.

57. D. Quillen. Cohomology of groups. Actes, Congrès intern. Math. Nice 1970. Tome 2, 47-51.

58. D. Quillen. The Adams conjecture. Topology 10(1971), 67-80.

59. D. Quillen. On the cohomology and K-theory of the general linear groups over a finite field. Annals Math. 96(1972), 552-586.

60. D. Quillen. Letter to J. Milnor, dated July 26, 1972.

61. D. Quillen. Higher K-theory for categories with exact sequences. London Math. Soc. Lecture Note Series 11, 1974, 95-103.

62. D. Quillen. Higher algebraic K-theory I. Lecture Notes in Mathematics Vol. 341, 85-147. Springer, 1973.

63. F. Quinn. Surgery on Poincaré and normal spaces. Bull. Amer. Math. Soc. 78(1972), 262-268.

64. F. Quinn. Geometric surgery II, bordism and bundle theories. In preparation.

65. N. Ray. The symplectic J-homomorphism. Invent. Math. 12(1971), 237-248.

66. N. Ray. Bordism J-homomorphisms. Illinois J. Math. 18(1974), 290-309.

67. G. Segal. Configuration spaces and iterated loop spaces. Invent. Math. 21(1973), 213-221.

68. G. Segal. Categories and cohomology theories. Topology 13(1974), 293-312.

69. J.P. Serre. Représentations linéaires des groupes finis. Hermann, 1967.

70. V.P. Snaith. Dyer-Lashof operations in K-theory, part I. Lecture Notes in Mathematics. Vol. 496, 103-294. Springer, 1975.

71. D. Sullivan. Geometric topology seminar. Mimeographed notes. Princeton.

72. D. Sullivan. Geometric topology, Part I. Localization, periodicity, and Galois symmetry. Mimeographed Notes. M.I.T.

73. D. Sullivan. Genetics of homotopy theory and the Adams conjecture. Annals Math. 100(1974), 1-79.

74. L.R. Taylor. Observations on orientability. Preprint.

75. J. Tornehave. Delooping the Quillen map. Thesis. M.I.T. 1971.

76. J. Tornehave. On BSG and the symmetric groups. Preprint.

77. J. Tornehave. The splitting of spherical fibration theory at odd primes. Preprint.

78. A. Tsuchiya. Characteristic classes for PL micro bundles. Nagoya Math. J. 43(1971), 169-198.

79. J.B. Wagoner. Delooping classifying spaces in algebraic K-theory. Topology 11(1972), 349-370.

80. G.W. Whitehead. Generalized homology theories. Trans. Amer. Math. Soc. 103(1962), 227-283.

Index

Vol. 460: O. Loos, Jordan Pairs. XVI, 218 pages. 1975.

Vol. 461: Computational Mechanics. Proceedings 1974. Edited by J. T. Oden. VII, 328 pages. 1975.

Vol. 462: P. Gérardin, Construction de Séries Discrètes p-adiques. »Sur les séries discrètes non ramifiées des groupes réductifs déployés p-adiques«. III, 180 pages. 1975.

Vol. 463: H.-H. Kuo, Gaussian Measures in Banach Spaces. VI, 224 pages. 1975.

Vol. 464: C. Rockland, Hypoellipticity and Eigenvalue Asymptotics. III, 171 pages. 1975.

Vol. 465: Séminaire de Probabilités IX. Proceedings 1973/74. Edité par P. A. Meyer. IV, 589 pages. 1975.

Vol. 466: Non-Commutative Harmonic Analysis. Proceedings 1974. Edited by J. Carmona, J. Dixmier and M. Vergne. VI, 231 pages. 1975.

Vol. 467: M. R. Essén, The Cos $\pi\lambda$ Theorem. With a paper by Christer Borell. VII, 112 pages. 1975.

Vol. 468: Dynamical Systems – Warwick 1974. Proceedings 1973/74. Edited by A. Manning. X, 405 pages. 1975.

ol. 469: E. Binz, Continuous Convergence on C(X). IX, 140 pages. 975.

ol. 470: R. Bowen, Equilibrium States and the Ergodic Theory of nosov Diffeomorphisms. III, 108 pages. 1975.

ol. 471: R. S. Hamilton, Harmonic Maps of Manifolds with Boundary. 168 pages. 1975.

l. 472: Probability-Winter School. Proceedings 1975. Edited by Ciesielski, K. Urbanik, and W. A. Woyczyński. VI, 283 pages. 75.

l. 473: D. Burghelea, R. Lashof, and M. Rothenberg, Groups of tomorphisms of Manifolds. (with an appendix by E. Pedersen) 156 pages. 1975.

l. 474: Séminaire Pierre Lelong (Analyse) Année 1973/74. Edité P. Lelong. VI, 182 pages. 1975.

. 475: Répartition Modulo 1. Actes du Colloque de Marseille-miny, 4 au 7 Juin 1974. Edité par G. Rauzy. V, 258 pages. 1975. 75.

. 476: Modular Functions of One Variable IV. Proceedings 1972. ited by B. J. Birch and W. Kuyk. V, 151 pages. 1975.

. 477: Optimization and Optimal Control. Proceedings 1974. ted by R. Bulirsch, W. Oettli, and J. Stoer. VII, 294 pages. 1975.

478: G. Schober, Univalent Functions – Selected Topics. V, pages. 1975.

479: S. D. Fisher and J. W. Jerome, Minimum Norm Extremals Function Spaces. With Applications to Classical and Modern lysis. VIII, 209 pages. 1975.

480: X. M. Fernique, J. P. Conze et J. Gani, Ecole d'Eté de babilités de Saint-Flour IV-1974. Edité par P.-L. Hennequin. 93 pages. 1975.

481: M. de Guzmán, Differentiation of Integrals in R^n. XII, 226 es. 1975.

482: Fonctions de Plusieurs Variables Complexes II. Séminaire çois Norguet 1974–1975. IX, 367 pages. 1975.

483: R. D. M. Accola, Riemann Surfaces, Theta Functions, and ian Automorphisms Groups. III, 105 pages. 1975.

484: Differential Topology and Geometry. Proceedings 1974. d by G. P. Joubert, R. P. Moussu, and R. H. Roussarie. IX, pages. 1975.

485: J. Diestel, Geometry of Banach Spaces – Selected Topics. 2 pages. 1975.

486: S. Stratila and D. Voiculescu, Representations of AF-bras and of the Group U (∞). IX, 169 pages. 1975.

87: H. M. Reimann und T. Rychener, Funktionen beschränkter rer Oszillation. VI, 141 Seiten. 1975.

88: Representations of Algebras, Ottawa 1974. Proceedings Edited by V. Dlab and P. Gabriel. XII, 378 pages. 1975.

Vol. 489: J. Bair and R. Fourneau, Etude Géométrique des Espaces Vectoriels. Une Introduction. VII, 185 pages. 1975.

Vol. 490: The Geometry of Metric and Linear Spaces. Proceedings 1974. Edited by L. M. Kelly. X, 244 pages. 1975.

Vol. 491: K. A. Broughan, Invariants for Real-Generated Uniform Topological and Algebraic Categories. X, 197 pages. 1975.

Vol. 492: Infinitary Logic: In Memoriam Carol Karp. Edited by D. W. Kueker. VI, 206 pages. 1975.

Vol. 493: F. W. Kamber and P. Tondeur, Foliated Bundles and Characteristic Classes. XIII, 208 pages. 1975.

Vol. 494: A Cornea and G. Licea. Order and Potential Resolvent Families of Kernels. IV, 154 pages. 1975.

Vol. 495: A. Kerber, Representations of Permutation Groups II. V, 175 pages. 1975.

Vol. 496: L. H. Hodgkin and V. P. Snaith, Topics in K-Theory. Two Independent Contributions. III, 294 pages. 1975.

Vol. 497: Analyse Harmonique sur les Groupes de Lie. Proceedings 1973–75. Edité par P. Eymard et al. VI, 710 pages. 1975.

Vol. 498: Model Theory and Algebra. A Memorial Tribute to Abraham Robinson. Edited by D. H. Saracino and V. B. Weispfenning. X, 463 pages. 1975.

Vol. 499: Logic Conference, Kiel 1974. Proceedings. Edited by G. H. Müller, A. Oberschelp, and K. Potthoff. V, 651 pages 1975.

Vol. 500: Proof Theory Symposion, Kiel 1974. Proceedings. Edited by J. Diller and G. H. Müller. VIII, 383 pages. 1975.

Vol. 501: Spline Functions, Karlsruhe 1975. Proceedings. Edited by K. Böhmer, G. Meinardus, and W. Schempp. VI, 421 pages. 1976.

Vol. 502: János Galambos, Representations of Real Numbers by Infinite Series. VI, 146 pages. 1976.

Vol. 503: Applications of Methods of Functional Analysis to Problems in Mechanics. Proceedings 1975. Edited by P. Germain and B. Nayroles. XIX, 531 pages. 1976.

Vol. 504: S. Lang and H. F. Trotter, Frobenius Distributions in GL_2-Extensions. III, 274 pages. 1976.

Vol. 505: Advances in Complex Function Theory. Proceedings 1973/74. Edited by W. E. Kirwan and L. Zalcman. VIII, 203 pages. 1976.

Vol. 506: Numerical Analysis, Dundee 1975. Proceedings. Edited by G. A. Watson. X, 201 pages. 1976.

Vol. 507: M. C. Reed, Abstract Non-Linear Wave Equations. VI, 128 pages. 1976.

Vol. 508: E. Seneta, Regularly Varying Functions. V, 112 pages. 1976.

Vol. 509: D. E. Blair, Contact Manifolds in Riemannian Geometry. VI, 146 pages. 1976.

Vol. 510: V. Poènaru, Singularités C^∞ en Présence de Symétrie. V, 174 pages. 1976.

Vol. 511: Séminaire de Probabilités X. Proceedings 1974/75. Edité par P. A. Meyer. VI, 593 pages. 1976.

Vol. 512: Spaces of Analytic Functions, Kristiansand, Norway 1975. Proceedings. Edited by O. B. Bekken, B. K. Øksendal, and A. Stray. VIII, 204 pages. 1976.

Vol. 513: R. B. Warfield, Jr. Nilpotent Groups. VIII, 115 pages. 1976.

Vol. 514: Séminaire Bourbaki vol. 1974/75. Exposés 453 – 470. IV, 276 pages. 1976.

Vol. 515: Bäcklund Transformations. Nashville, Tennessee 1974. Proceedings. Edited by R. M. Miura. VIII, 295 pages. 1976.

Vol. 516: M. L. Silverstein, Boundary Theory for Symmetric Markov Processes. XVI, 314 pages. 1976.

Vol. 517: S. Glasner, Proximal Flows. VIII, 153 pages. 1976.

Vol. 518: Séminaire de Théorie du Potentiel, Proceedings Paris 1972–1974. Edité par F. Hirsch et G. Mokobodzki. VI, 275 pages. 1976.

Vol. 519: J. Schmets, Espaces de Fonctions Continues. XII, 150 pages. 1976.

Vol. 520: R. H. Farrell, Techniques of Multivariate Calculation. X, 337 pages. 1976.

Vol. 521: G. Cherlin, Model Theoretic Algebra – Selected Topics. IV, 234 pages. 1976.

Vol. 522: C. O. Bloom and N. D. Kazarinoff, Short Wave Radiation Problems in Inhomogeneous Media: Asymptotic Solutions. V. 104 pages. 1976.

Vol. 523: S. A. Albeverio and R. J. Høegh-Krohn, Mathematical Theory of Feynman Path Integrals. IV, 139 pages. 1976.

Vol. 524: Séminaire Pierre Lelong (Analyse) Année 1974/75. Edité par P. Lelong. V, 222 pages. 1976.

Vol. 525: Structural Stability, the Theory of Catastrophes, and Applications in the Sciences. Proceedings 1975. Edited by P. Hilton. VI, 408 pages. 1976.

Vol. 526: Probability in Banach Spaces. Proceedings 1975. Edited by A. Beck. VI, 290 pages. 1976.

Vol. 527: M. Denker, Ch. Grillenberger, and K. Sigmund, Ergodic Theory on Compact Spaces. IV, 360 pages. 1976.

Vol. 528: J. E. Humphreys, Ordinary and Modular Representations of Chevalley Groups. III, 127 pages. 1976.

Vol. 529: J. Grandell, Doubly Stochastic Poisson Processes. X, 234 pages. 1976.

Vol. 530: S. S. Gelbart, Weil's Representation and the Spectrum of the Metaplectic Group. VII, 140 pages. 1976.

Vol. 531: Y.-C. Wong, The Topology of Uniform Convergence on Order-Bounded Sets. VI, 163 pages. 1976.

Vol. 532: Théorie Ergodique. Proceedings 1973/1974. Edité par J.-P. Conze and M. S. Keane. VIII, 227 pages. 1976.

Vol. 533: F. R. Cohen, T. J. Lada, and J. P. May, The Homology of Iterated Loop Spaces. IX, 490 pages. 1976.

Vol. 534: C. Preston, Random Fields. V, 200 pages. 1976.

Vol. 535: Singularités d'Applications Differentiables. Plans-sur-Bex. 1975. Edité par O. Burlet et F. Ronga. V, 253 pages. 1976.

Vol. 536: W. M. Schmidt, Equations over Finite Fields. An Elementary Approach. IX, 267 pages. 1976.

Vol. 537: Set Theory and Hierarchy Theory. Bierutowice, Poland 1975. A Memorial Tribute to Andrzej Mostowski. Edited by W. Marek, M. Srebrny and A. Zarach. XIII, 345 pages. 1976.

Vol. 538: G. Fischer, Complex Analytic Geometry. VII, 201 pages. 1976.

Vol. 539: A. Badrikian, J. F. C. Kingman et J. Kuelbs, Ecole d'Eté de Probabilités de Saint Flour V-1975. Edité par P.-L. Hennequin. IX, 314 pages. 1976.

Vol. 540: Categorical Topology, Proceedings 1975. Edited by E. Binz and H. Herrlich. XV, 719 pages. 1976.

Vol. 541: Measure Theory, Oberwolfach 1975. Proceedings. Edited by A. Bellow and D. Kölzow. XIV, 430 pages. 1976.

Vol. 542: D. A. Edwards and H. M. Hastings, Čech and Steenrod Homotopy Theories with Applications to Geometric Topology. VII, 296 pages. 1976.

Vol. 543: Nonlinear Operators and the Calculus of Variations, Bruxelles 1975. Edited by J. P. Gossez, E. J. Lami Dozo, J. Mawhin, and L. Waelbroeck, VII, 237 pages. 1976.

Vol. 544: Robert P. Langlands, On the Functional Equations Satisfied by Eisenstein Series. VII, 337 pages. 1976.

Vol. 545: Noncommutative Ring Theory. Kent State 1975. Edited by J. H. Cozzens and F. L. Sandomierski. V, 212 pages. 1976.

Vol. 546: K. Mahler, Lectures on Transcendental Numbers. Edited and Completed by B. Diviš and W. J. Le Veque. XXI, 254 pages. 1976.

Vol. 547: A. Mukherjea and N. A. Tserpes, Measures on Topological Semigroups: Convolution Products and Random Walks. V, 197 pages. 1976.

Vol. 548: D. A. Hejhal, The Selberg Trace Formula for PSL (2,$\mathbb{R}$). Volume I. VI, 516 pages. 1976.

Vol. 549: Brauer Groups, Evanston 1975. Proceedings. Edited by D. Zelinsky. V, 187 pages. 1976.

Vol. 550: Proceedings of the Third Japan – USSR Symposium on Probability Theory. Edited by G. Maruyama and J. V. Prokhorov. VI, 722 pages. 1976.

Vol. 551: Algebraic K-Theory, Evanston 1976. Proceedings. Edited by M. R. Stein. XI, 409 pages. 1976.

Vol. 552: C. G. Gibson, K. Wirthmüller, A. A. du Plessis and E. J. N. Looijenga. Topological Stability of Smooth Mappings. V, 155 pages. 1976.

Vol. 553: M. Petrich, Categories of Algebraic Systems. Vector and Projective Spaces, Semigroups, Rings and Lattices. VIII, 217 pages. 1976.

Vol. 554: J. D. H. Smith, Mal'cev Varieties. VIII, 158 pages. 1976.

Vol. 555: M. Ishida, The Genus Fields of Algebraic Number Fields. VII, 116 pages. 1976.

Vol. 556: Approximation Theory. Bonn 1976. Proceedings. Edited by R. Schaback and K. Scherer. VII, 466 pages. 1976.

Vol. 557: W. Iberkleid and T. Petrie, Smooth S^1 Manifolds. III, 163 pages. 1976.

Vol. 558: B. Weisfeiler, On Construction and Identification of Graphs. XIV, 237 pages. 1976.

Vol. 559: J.-P. Caubet, Le Mouvement Brownien Relativiste. IX, 212 pages. 1976.

Vol. 560: Combinatorial Mathematics, IV, Proceedings 1975. Edite[d] by L. R. A. Casse and W. D. Wallis. VII, 249 pages. 1976.

Vol. 561: Function Theoretic Methods for Partial Differential Equation[s] Darmstadt 1976. Proceedings. Edited by V. E. Meister, N. Wec[k] and W. L. Wendland. XVIII, 520 pages. 1976.

Vol. 562: R. W. Goodman, Nilpotent Lie Groups: Structure an[d] Applications to Analysis. X, 210 pages. 1976.

Vol. 563: Séminaire de Théorie du Potentiel. Paris, No. 2. Proceeding[s] 1975–1976. Edited by F. Hirsch and G. Mokobodzki. VI, 292 page[s] 1976.

Vol. 564: Ordinary and Partial Differential Equations, Dundee 197[6] Proceedings. Edited by W. N. Everitt and B. D. Sleeman. XVIII, 5[…] pages. 1976.

Vol. 565: Turbulence and Navier Stokes Equations. Proceeding[s] 1975. Edited by R. Temam. IX, 194 pages. 1976.

Vol. 566: Empirical Distributions and Processes. Oberwolfach 197[6] Proceedings. Edited by P. Gaenssler and P. Révész. VII, 146 page[s] 1976.

Vol. 567: Séminaire Bourbaki vol. 1975/76. Exposés 471–488. [IV,] 303 pages. 1977.

Vol. 568: R. E. Gaines and J. L. Mawhin, Coincidence Degree, a[nd] Nonlinear Differential Equations. V, 262 pages. 1977.

Vol. 569: Cohomologie Etale SGA 4½. Séminaire de Géomét[rie] Algébrique du Bois-Marie. Edité par P. Deligne. V, 312 pages. 19[77].

Vol. 570: Differential Geometrical Methods in Mathematical Physi[cs,] Bonn 1975. Proceedings. Edited by K. Bleuler and A. Reetz. V[…] 576 pages. 1977.

Vol. 571: Constructive Theory of Functions of Several Variab[les,] Oberwolfach 1976. Proceedings. Edited by W. Schempp and K. Z[el]ler. VI, 290 pages. 1977

Vol. 572: Sparse Matrix Techniques, Copenhagen 1976. Edite[d by] V. A. Barker. V, 184 pages. 1977.

Vol. 573: Group Theory, Canberra 1975. Proceedings. Edite[d by] R. A. Bryce, J. Cossey and M. F. Newman. VII, 146 pages. 19[77].

Vol. 574: J. Moldestad, Computations in Higher Types. IV, [203] pages. 1977.

Vol. 575: K-Theory and Operator Algebras, Athens, Georgia 19[75.] Edited by B. B. Morrel and I. M. Singer. VI, 191 pages. 1977.

Vol. 576: V. S. Varadarajan, Harmonic Analysis on the Real [Re]ductive Groups. VI, 521 pages. 1977.

Vol. 577: J. P. May, E_∞ Ring Spaces and E_∞ Ring Spectra[.] 268 pages. 1977.